C000179830

THE DESIGN OF MODERN STEEL BRIDGES

SUKHEN CHATTERJEE

BE, MSc, DIC, PhD, MICE, MIStructE

OXFORD

BSP PROFESSIONAL BOOKS

LONDON   EDINBURGH   BOSTON

MELBOURNE   PARIS   BERIN   VIENNA

Copyright © Sukhen Chatterjee 1991

BSP Professional Books
A division of Blackwell Scientific
  Publications Ltd
Editorial offices:
Osney Mead, Oxford OX2 0EL
25 John Street, London WC1N 2BL
23 Ainslie Place, Edinburgh EH3 6AJ
3 Cambridge Center, Cambridge,
  MA 02142, USA
54 University Street, Carlton,
  Victoria 3053, Australia

All rights reserved. No part of this
publication may be reproduced, stored
in a retrieval system, or transmitted
in any form or by any means, electronic,
mechanical, photocopying, recording
or otherwise without the prior
permission of the publisher.

First published 1991

Set by Setrite Typesetters Ltd
Printed and bound in Great Britain by
Hartnolls, Bodmin, Cornwall

DISTRIBUTORS

Marston Book Services Ltd
PO Box 87
Oxford OX2 0DT
(*Orders*: Tel: 0865 791155
        Fax: 0865 791927
        Telex: 837515)

USA
  Blackwell Scientific Publications, Inc.
  3 Cambridge Center
  Cambridge, MA 02142
  (*Orders*: Tel: (800) 759−6102)

Canada
  Oxford University Press
  70 Wynford Drive
  Don Mills
  Ontario M3C 1J9
  (*Orders*: Tel: (416) 441−2941)

Australia
  Blackwell Scientific Publications
  (Australia) Pty Ltd
  54 University Street
  Carlton, Victoria 3053
  (*Orders*: Tel: (03) 347−0300)

British Library
Cataloguing in Publication Data
Chatterjee, Sukhen
  The design of modern steel bridges.
  1. Steel bridges. Design
  I. Title
  624′.25

ISBN 0−632−01829−1

Contents

Preface

Bridges are great symbols of mankind's conquest of space. The sight of the crimson tracery of the Golden Gate Bridge against a setting sun in the Pacific Ocean, or the arch of the Garabit Viaduct soaring triumphantly above the deep gorge, fills one's heart with wonder and admiration for the art of their builders. They are the enduring expressions of mankind's determination to remove all barriers in its pursuit of a better and freer world. Their design and building schemes are conceived in dream-like visions. But vision and determination are not enough. All the physical forces of nature and gravity must be understood with mathematical precision and such forces have to be resisted by manipulating the right materials in the right pattern. This requires both the inspiration of an artist and the skill of an artisan.

Scientific knowledge about materials and structural behaviour has expanded tremendously, and computing techniques are now widely available to manipulate complex theories in innumerable ways very quickly. But it is still not possible to accurately cater for all the known and unknown intricacies. Even the most advanced theories and techniques have their approximations and exceptions. The wiser the scientist, the more he knows of his limitations. Hence scientific knowledge has to be tempered with a judgment as to how far to rely on mathematical answers and then what provision to make for the unknown realities. Great bridge-builders like Stephenson and Roebling provided practical solutions to some very complex structural problems, for which correct mathematical solutions were derived many years later; in fact the clue to the latter was provided by the former.

Great intuition and judgment spring from genius, but they can be helped along the way by an understanding of the mathematical theories. The object of this book is to explain firstly the nature of the problems associated with the building of bridges with steel as the basic material, and then the theories that are available to tackle them. The reader is assumed to have the basic degree-level knowledge of civil engineering, i.e. he may be a final-year undergraduate doing a project with bridges, or a qualified engineer entering into the field of designing and building steel bridges.

v

The book sets out with a technological history of the gradual development of different types of iron and steel bridges. A knowledge of this evolution from the earliest cast-iron ribbed arch, through the daring suspension and arch structures, on to the modern elegant plated spans, will contribute to a proper appreciation of the state-of-the-art today.

The basic properties of steel as a building material, and the successive improvement achieved by the metallurgist at the behest of the bridge-builder, are then described. The natural and the traffic-induced forces and phenomena that the bridge structure must resist are then identified and quantified with reference to the practices in different countries. This is followed by an explanation of the philosophy behind the process of the structural design of bridges, i.e. the basic functional aims and how the mathematical theories are applied to achieve them in spite of the unavoidable uncertainties inherent in natural forces, in idealised theories and in the construction processes. This subject is treated in the context of limit state and statistical probability concepts. Then follows detailed guidance on the design of plate and box girder bridges, the most common form of construction adopted for steel bridges in modern times. The buckling behaviour of various components, the effects of geometrical imperfections and large-deflection behaviour, and the phenomenon of post-buckling reserves are described in great detail. The rationale behind the requirements of various national codes and the research that helped their evolution are explained, and a few design examples are worked out to illustrate their intended use.

Sukhen Chatterjee

Acknowledgements

The plates in Chapter 1 are reproduced with kind permission of the following:

Flint & Neill Partnership, London — for Plates 3, 7, 12, 13, 14, 18, 19, 26, 27 and 33.

Steel Construction Institute, Ascot — for Plates 5, 17, 22, 24, 25, 29 and 31 (from the collection of the late Bernard Godfrey).

Acer, Freeman Fox, Guildford — for Plates 10, 11, 16, 21, 28 and 30.

Rendel, Palmer & Tritton, London — for Plates 4, 8 and 9.

Professor J. Harding of Surrey University — for Plate 20.

Mr B. Oakhill of BCSA, London — for Plate 23.

Travers Morgan, East Grinstead — for Plate 32.

Steinman, Boynton, Gronquist & London of New York — Plate 15.

Ironbridge Gorge Museum Trust — for Plate 1.

GAF Corporation — for Plate 6.

The front cover design is from a photograph of the motorway interchange of the M4 and M25 in England, courtesy of Steel Construction Institute.

Chapter 1
Types and History of Steel Bridges

1.1 Bridge types

There are five basic types of steel bridges:

(1) Girder bridges − flexure or bending between vertical supports is the main structural action in this type. They may be further sub-divided into simple spans, continuous spans and suspended-and-cantilevered spans as illustrated in Fig. 1.1.

(2) Rigid frame bridges − in this type the longitudinal girders are made structurally continuous with the vertical or inclined supporting members by means of moment-carrying joints; flexure with some axial force is the main structural action in this type.

(3) Arches − in which the loads are transferred to the foundations by arches as the main structural element; axial compression in the arch rib is the main structural action, combined with some bending. The horizontal thrust at the ends is resisted either by the foundations or by a tie running longitudinally for the full span length; the latter type is called a tied or a bow-string arch.

(4) Cable-stayed bridges − in which the main longitudinal girders are supported by a few or many ties in the vertical or near-vertical plane, which are hung from one or more tall towers and are usually anchored at the bottom to the girders.

(5) Suspension bridges − in which the bridge deck is suspended from cables stretched over the gap to be bridged, anchored to the ground at two ends and passing over tall towers erected at or near the two edges of the gap.

The first three types and the deck structure of the last two types of bridges may be either solid-web girders or truss (or lattice) girders.

1

Arch bridges

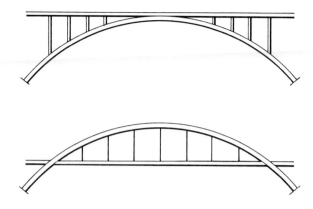

Cable-stayed bridges

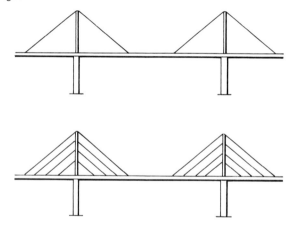

Suspension bridge

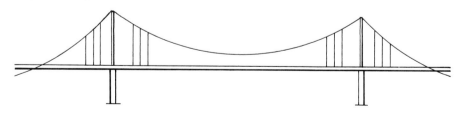

Fig. 1.1   Different types of bridges.

Girder Bridges

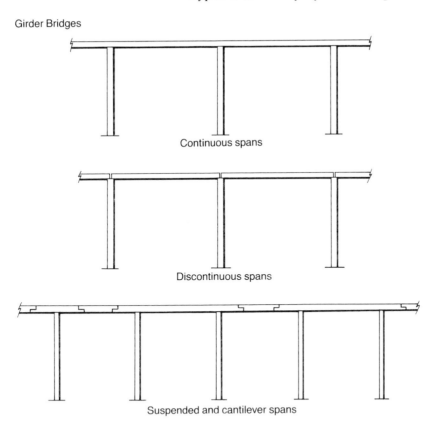

Continuous spans

Discontinuous spans

Suspended and cantilever spans

Rigid Frame Bridges

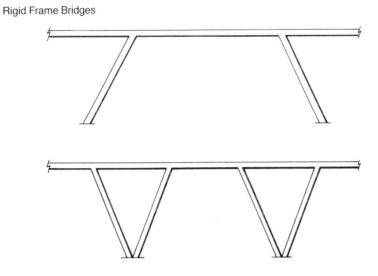

Fig. 1.1   (contd)

1.2 History of bridges

1.2.1 Iron bridges

Iron was used in Europe for building cannons and machinery in the 16th century, but it was not until the late 18th century, in the wake of the first industrial revolution, that iron was first used for structures. The world's first iron bridge was the famous Coalbrookdale bridge in the county of Shropshire in England, spanning over the 100 ft (30.5 m) width of the River Severn, designed by Thomas Pritchard and built by ironmasters Darby and Wilkinson in 1777–79. It was made of a series of semicircular cast-iron arch ribs side by side; in each vertical plane the bottom arch rib was continuous over the span, stiffened by two upper ribs that terminated at and propped the road level but were not otherwise continuous over the span. The quality and workmanship of the 400 ton ironwork were such that the bridge is standing even today, after over 200 years, though not carrying today's vehicles.

Coalbrookdale iron bridge was, however, built with concepts that are traditional with stone bridges, e.g. a semicircular shape and spandrel built

Plate 1.   Coalbrookdale iron bridge in Shropshire, England (1777–79).

Plate 2. Suspension bridge across the Menai Straits, Wales (1819–26).

with tiers of ribs. Thomas Telford recognised that the special properties of iron, e.g. its considerably lighter weight and higher strength, would permit longer and flatter arches. In 1796 he built the Buildwas bridge over the Severn in Shropshire in cast-iron, a 130 ft (40 m) span arc segment.

Earlier the famous American humanist Tom Paine designed a 400 ft (122 m) span cast-iron bridge over the Schuylkill in Philadelphia, ordering the ironwork from Yorkshire, England. However, the project was delayed and the iron was used to build a 236 ft (72 m) span bridge over the Wear in Sunderland simultaneously with Buildwas. These bridges led the way to many more iron bridges in the first two decades of the 19th century in England and France, the most notable being the Vauxhall and Southwark bridges over the Thames in London (each using over 6000 tons of iron) and Pont du Louvre and Pont d'Austerlitz over the Seine in Paris (the latter has since been replaced). In the early days cast-iron was slotted and dovetailed like timber construction before bolting was discovered.

In 1814 Thomas Telford proposed a suspension bridge with cables made of flat wrought iron links to cross the Mersey at Runcorn — a main span of 1000 ft (305 m) and two side spans of 500 ft! The suspension principle has been used for building pedestrian bridges in India, China and South

Plate 3.   Arch Bridge over Oxford Canal, England (1832–34).

Plate 4.   Chelsea Bridge over Thames, England (1851–58).

America since time immemorial; they were supported by bundles of vines or osiers, bamboo strips, plaited ropes, etc., and sometimes even had plank floors and hand rails. Telford, collaborating with Samuel Brown, made experiments with wrought iron and decided that cables made of wrought iron eyebar chains could be used with a working stress of 5 tons per square inch (77 N/mm^2), compared with only 1.25 tons/in^2 tensile working stress of cast-iron. The Mersey bridge did not materialise. The Holyhead Road, however, proposed for improved communication between Britain and Ireland, required a bridge over the Menai Straits, and Telford proposed in 1817 a suspension bridge of 580 ft (177 m) main span. Work on site started in 1819, and in 1826 the world's first iron suspension bridge for vehicles was completed. This was the world's first bridge over sea water. The bridge had 100 ft (30.5 m) clearance over the high water of the Irish Sea and took 2000 tons of wrought iron (compared with 6000 tons of iron for the Vauxhall arch bridge). It had no stiffening girder and no wind bracing; its deck had to be replaced in 1839, 1893 and again in 1939. Telford also built the 327 ft (100 m) span suspension bridge at Conway for the same Holyhead Road at about the same time. The success with these two suspension bridges brought about a new era of long-span bridges.

Plate 5.   Albert Bridge over Thames, England (1864—73).

Plate 6. Brooklyn Bridge, New York (1867–83).

Isambard Kingdom Brunel built Hungerford pedestrian suspension bridge over the Thames at Charing Cross which, however, had to be removed 20 years later to make room for the present railway bridge. William Clark built Hammersmith Bridge, Norfolk Bridge at Shoreham, the bridge at Marlow over the Thames and the 666 ft (203 m) span bridge over the Danube at Budapest — all suspension bridges. Several other suspension bridges with wire cables were built in Europe, the most remarkable being the Grand Pont at Fribourg, Switzerland, by Chaley, which had a 800 ft (244 m) span supported by four cables each of 1056 wires, 3 mm diameter. A competition was held for the design of a bridge over the Avon Gorge at Clifton, Bristol. Brunel submitted a design for a suspension bridge of 1160 ft (354 m) span. Telford was a judge for the competition and did not consider such a long span practicable. In a second competition in 1850, Brunel's design of a 600 ft (183 m) span was accepted. Work started but was abandoned due to the contractor going bankrupt. In 1860, a year after Brunel's death, work was resumed with some changes in the design and completed in 1864; the chains of Brunel's Hungerford bridge were reused here — they had wrought iron shafts with eyes welded to their ends by hot-hammering. This beautiful bridge is still carrying vehicles — a great testimony to a very great engineer.

By the middle of the 19th century, good quality wrought iron was being produced commercially, replacing cast-iron for structural work and being used extensively for shipbuilding. This material was ductile, malleable, strong in tension and could be riveted. William Fairbairn had already designed a riveting machine.

In the second half of the 18th century, the coal industry in England was using steam engines for pumping out water, and wooden or iron rails for moving coal wagons. In the first decade of the 19th century, several collieries around Newcastle had steam boilers on wheels running on rails by means of ratchet wheels for hauling coal wagons. In 1814 George Stephenson built an engine which did not need any ratchets to run on

Plate 7.   Forth Railway Bridge, Scotland (1881—90).

iron rails. In 1825 Stephenson's Rocket engine ran on the Stockport-to-Darlington railway. This railway was followed by Manchester–Liverpool and London–Birmingham railways. Soon railways grew all over Britain, then in Europe and North America. This produced an insatiable demand for bridges (and tunnels), but these bridges had to be sturdy enough to carry not only the heavy weight of the locomotives, but also their severe pounding on the rails. They also had to be built on a nearly level grade; otherwise the locomotives could not pull the wagons up.

George Stephenson built two types of bridges for his railways – a simple beam of cast-iron for short spans over roads and canals, and cast-iron arches for longer spans. The most striking example of the latter type was the Newcastle High Level Bridge; his son Robert played a significant part in its design and construction. The bridge consisted of six bow-string arches, each with a horizontal tie between the springing points to resist the end thrust, with the railway on the top and the road suspended underneath by wrought iron rods 120 ft (37 m) above water. It was completed in August 1849 and a few days after the opening Queen Victoria stopped her train on it to admire the view.

Plate 8. Howrah Bridge, Calcutta, India.

Robert Stephenson was already considering how to cross the Menai Straits and the Conway river for his Chester−Holyhead Railway. Suspension bridges built up to then to carry horse-drawn carriages exhibited a lack of rigidity and a weakness in windy conditions, and hence could not possibly withstand the heavy and rhythmic pounding of locomotives. Several such bridges carrying roads had either fallen down or suffered great damage; for example, the one at Broughton had collapsed under a column of marching soldiers and the chain-pier bridge at Brighton had been blown down by a storm. This list also included bridges at Tweed, Nassau in Germany, Roche Bernard in France and several in America.

The first suspension bridge to carry a railway was built by Samuel Brown in 1830 over the Tees; it sagged when a train came over it, and the engine could not climb up the steep gradient that the deflection of the structure formed ahead of it. Robert Stephenson decided that Telford's road bridge solution of a suspension bridge would not be appropriate to carry a railway over the Menai Straits, nor could a cast-iron arch be built here, as the Admiralty would permit neither the reduction in headroom near the springing points of the arch construction nor the temporary navigational blockage that the timber centring would cause. Stephenson had already decided that a rocky island in the Menai channel called the Britannia Rock would support an intermediate pier.

Stephenson hit upon the idea of two massive wrought iron tubes through which the trains could run. At his request William Fairbairn conducted tests on circular, rectangular and elliptical shapes, and also on wrought iron stiffened and cellular panels for their compressive strength. The Conway crossing was ready first, and in 1848 two huge tubes 400 ft (122 m) long were floated out on pontoons, lifted up and placed in their correct positions during a falling tide. A year later, in 1849, the four tubes of the Menai crossing, two 460 ft (140 m) and two 230 ft (70 m) spans, were similarly erected. As box girder bridges, they were highly ingenious and unique for many decades; they were also the giant forerunners of thousands of plate girder bridges that became the most popular type of bridge construction all over the world. The Britannia Bridge at Menai was severely damaged by a fire in May 1970 and had to be rebuilt in the shape of a spandrel braced arch as originally proposed by Rennie and Telford. A roadway was also added on an upper level.

In the United States, railroad construction started in the early 1830s. The early railway bridges were mostly patented truss types ('Howe', 'Pratt', 'Warren', etc.) with wooden compression members and wrought iron tension members. These were followed by a composite truss system of cast-iron compression members and wrought iron tension members. In 1842 Charles Ellet built a suspension bridge over Schuylkill river at Fairmont, Pennsylvania, to replace Lewis Wernwag's 340 ft (104 m) span

Plate 9.   Hardinge Bridge, India.

Colossus Bridge destroyed by fire. The latter was a timber bridge formed in the shape of a gently curved arch reinforced by trusses the diagonals of which were iron rods — the first use of iron in a long-span bridge in America. Ellet's suspended span was supported by ten wire cables. In 1848 Ellet started to build the first ever bridge across the 800 ft (244 m) wide chasm below the Niagara Falls to carry a railway. To carry the first wire, he offered a prize of five dollars to fly a kite across. After the first wire cable was stretched in this way, the showman that he was, he hauled himself across the gorge in a wire basket at a height of 250 ft (76 m) above the swirling water! He then built a 7.5 ft (2.3 m) wide service bridge without railings, rode across on a horse and started collecting fares. Then he fell out with the promoters and withdrew, leading to the appointment of John Roebling to erect a new bridge. In 1841 Roebling had already patented his idea of forming cables from parallel wires bound into a compact bunch by binding wire.

In 1848 John Ellet had built another suspension bridge of 308 m (1010 ft) span over the Ohio river at Wheeling, West Virginia. In January 1854 it collapsed in a storm, due to aerodynamic vibration. Roebling realised that 'the destruction' of the Wheeling bridge was clearly 'owing to a want of stability, and not to a want of strength' — his own words. He also studied

Plate 10.   Volta Bridge, Africa.

Plate 11. Sydney Harbour Bridge, Australia.

the collapse of a suspension bridge in 1850 in Angers, France, under a marching regiment, and another in Licking, Kentucky, in 1854 under a drove of trotting cattle. His Grand Trunk bridge at Niagara had a 250 m (820 ft) span and had two decks, the upper one to carry a railway and the lower one a road; stiffening trusses 18 ft (5.5 m) deep of timber construction were provided between the two decks — the first stiffening girder used for a suspension bridge. The deck was supported by four main cables 10 inch (254 mm) in diameter consisting of parallel wrought iron wires, uniformly tensioned and compacted into a bunch with binding wire.

   This was the birth of the modern suspension bridge, which must be ranked as one of history's greatest inventions. The deck was also supported from the tower directly by 64 diagonal stays, and more stays were later added below the deck and anchored to the gorge sides. The bridge was completed in 1855. Roebling proved, contrary to Stephenson's prediction, that suspension bridges could carry railways and were more economical than the tubular girder construction used by the latter at Menai. In reality, however, not many railway suspension bridges were later built, but Roebling's Niagara bridge was the forebear of a great number of suspension bridges carrying roads. Its wooden deck was replaced by iron and the masonry towers by steel in 1881 and 1885, respectively, and finally the whole bridge was replaced in 1897. Two other bridges were built across the Niagara gorge. Serrell's road bridge of 1043 ft (318 m) span was built in 1851, stiffened by Roebling by stays in 1855 and destroyed by a storm in 1864 when the stays were left loose. At the site of the present Rainbow Bridge, Keefer built a bridge of 1268 ft (387 m) span in 1869, which was destroyed by a storm in 1889. John Roebling and his son Washington went on to build several more suspension bridges, the most notable being the ones at Pittsburgh and Cincinnati, and the Great Brooklyn Bridge in New York.

1.2.2   Steel bridges

In the second half of the 19th century steel was developed and started replacing cast-iron as a structural material. The technique of using compressed air to sink caissons for foundations below water was also developed. In 1855–59 Brunel built the Chepstow Bridge over the River Wye and the Saltash Bridge over the Tamar to carry railways. These were a combination of arch and suspension structures. A large wrought iron tube formed the upper chord shaped like an arch; the lower chord was a pair of suspension chains in catenary profile. The tube and the chains were braced together by diagonal ties and vertical struts. The first glimpse of lattice girder bridges can be seen in these designs. To carry railways over the Rhine in Germany, several bridges were built in the second half of the century, the most remarkable among them being:

(1)   Two bridges in Koln built in 1859, each with four spans of 338 ft (103 m) with multiple criss-cross lattice main girders 27.9 ft (8.5 m) deep
(2)   A bridge at Mainz built in 1882 with four spans of 344 ft (105 m), with a combined structural system of an arched top chord, a catenary bottom chord and a lattice in-filling between them, as in Brunel's Wye and Saltash bridges.

In America, the end of the Civil War and the spread of railway construction resulted in growing demands for building bridges. To connect the Illinois and the Union Pacific railways a bridge was needed over the 1500 ft (457 m) wide mighty Mississippi river at St Louis, for which James Eads was commissioned in 1867. The sandy river bed was subject to considerable shift and scour, and rock lay at varying depths between 50 and 150 ft (15–45 m). Swirling water rose 40 ft (12 m) in summer, and in winter 20 ft (6 m) thick chunks of ice hurtled down. Eads proposed to sink caissons down to rock level by compressed air – a technique already being used in Europe (by Brunel in Saltash, for example), but often at the cost of illness and fatality of the workmen. Eads also decided that a suspension bridge would

Plate 12. George Washington Bridge, New York.

not be stiff enough to carry railway loading; he proposed one 520 ft (159 m) and two 502 ft (153 m) spans of lattice arch construction with steel − the first use of the recently discovered material in a bridge. Bessemer had already converted iron to steel by adding carbon in 1856 and Siemens developed the open hearth process in 1867. But the problem was to produce the enormous quantity of this new material to a guaranteed and uniform quality rightly demanded by Eads, for example a minimum 'elastic limit'. Money was raised in America and Europe, which Eads visited to acquaint himself with the latest bridge building techniques. He designed

Plate 13.   Golden Gate Bridge, San Francisco.

chords of 18 inch (450 mm) diameter tubes made with $\frac{1}{4}$ inch (6 mm) thick steel plates. Each length of tube had wrought iron threaded end pieces shrunk-fit and they were screwed together by sleeve couplings. The two tubes were spaced 12 ft (3.7 m) apart vertically and braced together with diagonal members. The arch ribs were erected by cantilevering, with a series of temporary tie-back cables supported from temporary towers built over the piers − the first cantilever erection of a bridge superstructure. This method had to allow for the effects of temperature, the extension of the temporary cables and the compression of the arch rib, and one of the fund raising conditions was to have to close the first arch by 19 September 1873. This closure was just achieved, but the span had to be packed with ice at night in order to insert the closing piece in the final gap. The bridge carried two rail tracks on the bottom deck and a roadway on top, and is still in use. This bridge was the precursor of a glittering series of engineering achievements in America, which made it the most prosperous country in the world.

In the 1850s and 1860s in America many truss bridges were built for the railway lines, but many of them fell down. Buckling of compression members was the frequent cause of these failures. The worst disaster was the collapse of such a bridge 157 ft (48 m) long in Ashtabula, Ohio, on 29

December 1876, when during a snow storm a train fell down from it and 80 passengers died. Three years later, on 28 December 1879, 18 months after its completion, the Tay bridge in Scotland collapsed in a storm with 75 lives lost. Designed by Thomas Bouch, this 2 mile long bridge had 13 navigation spans of 245 ft (75 m), made of wrought iron trusses high above the water. In the subsequent inquiry it was established that the design did not allow for adequate horizontal wind loading. This was the first known example of a bridge failure due to the static horizontal pressure of wind drag, as opposed to the many failures of the early suspension bridges due to aerodynamic oscillations. The Tay Bridge was rebuilt, and all subsequent bridges were designed for a Board of Trade specified wind pressure of 56 lb per square foot (2.7 kN/m^2). In America, competitive supply of patented bridge types was subjected to a stricter regime of government regulations and independent supervision. Waddell led the movement for independent bridge design and supervision by consulting engineers; he himself was responsible for building hundreds of major bridges.

In 1867 John Roebling and his son Washington started to build Brooklyn bridge connecting Long Island with New York across East river. Its span of 1596 ft (487 m) nearly doubled the previous longest span built and it had to carry two railway lines, two tram-lines, a roadway and a footway.

Plate 14. Tamar Suspension Bridge, England (Brunel's bridge can also be seen).

Plate 15.   Mackinac Bridge, Michigan.

John Roebling died in 1869 due to an accident on the site, Washington completing the construction in 1883. Caissons were sunk by compressed air, amidst problems of 'caisson disease' (which crippled Washington himself), 'blowing' and fire. Galvanised cast steel wires of 16 000 lb/in^2 (110 N/mm^2) tensile strength were specified for the main cables; they were spun wire by wire by the then radical spinning method. To provide stability against wind forces and to supplement the capacity of the main cables, the suspended deck was held by diagonal cable stays radiating from the tower top. The graceful yet robust structure of Brooklyn Bridge was a landmark of human achievement, vision and determination.

The second half of the 19th century saw great advances in materials, machines and structural theories. Use of steel, banned in bridge construction in Britain by the Board of Trade till 1877, became common. Air compressors and hydraulic machines were developed for aiding construction. James Clerk Maxwell, Rankine and other engineering professors developed theories for analysis of suspension cables, lattice girders, bending moments and shear forces in beams, deflection calculations and buckling of struts. These developments and the unsuitability of suspension bridges for carrying railways, heralded the era of great trussed cantilever spans, led by the mighty Forth railway bridge. Designed in 1881 by John Fowler and Benjamin Baker, and construction completed in 1890 by Messrs Tancred, Arrol & Co, this bridge had two massive spans of 1710 ft (521 m), each consisting of two 680 ft (207 m) cantilevers and a 350 ft (107 m) suspended section. The depth of the truss at the piers was 350 ft (107 m). A German engineer called Gerber first developed the cantilever and suspended technique of bridge construction and quite a few such bridges were also built in America; it had the advantage of requiring no falsework over the gap. Projecting out in both directions, a cantilever structure was built on each pier and then a short suspended span was hung in between the tips of the two cantilevers. In the Forth Bridge, Baker built a third main pier on an island in the midstream, the bridge thus consisting of a triple cantilever with two suspended spans. The bridge carried two railway tracks 150 ft (46 m) above water. The specification for the steel required a minimum ultimate strength of 30 ton/in^2 (463 N/mm^2) for tensile members and 34 ton/in^2 (525 N/mm^2) for struts, and working stresses were a quarter of the ultimate strength. Over 50 000 tons of steel and 6 million rivets were used.

Steel truss bridges started going up all over the world. The Forth railway bridge in Scotland was followed by the Queensboro Bridge over the East River in New York which had two main spans of 1182 ft (360 m), a central span of 630 ft (192 m) and two anchor spans at the two shores – all made continuous in triangulated truss form, without any suspended spans of the Forth sort. This was followed by the start of construction in 1904 of the Quebec Bridge over the St Lawrence River in Canada which had a central span of 1800 ft (549 m). The bridge consisted of two giant truss cantilevers on two main piers, with a suspended span in the middle. The two anchor spans

Plate 16. Forth Road Bridge, Scotland.

Plate 17.   Salazar Bridge over the Tagus, Portugal.

were first built on falsework; then the cantilever arms on the river were erected member by member by cranes operating on the already erected structure. The two cantilevers having been completed on the two piers in this way, the members of the suspended spans were also being erected from both sides in this way when there were signs of buckling on the web plates of the compression chord members near the south pier and some rivets were found broken. Theodore Cooper, the respected elderly consulting engineer, who was not present on site, sent orders to stop erection, but work continued, and on 29 August 1907, the whole structure collapsed into the river, killing 75 men. In the subsequent inquiry and investigations it became clear that the lacing system and the splice joints of the compression members were not able to resist the effects of the buckling tendency of the compression members. In 1916 a new, slightly wider, structure was being rebuilt on new foundations; the two cantilevers had been completed and the entire 5000 ton suspended span, built on-shore and floated out, was being lifted up by hydraulic jacks. Then a casting support block at one corner failed, and the span slid off and fell into the water. The suspended span was rebuilt and erected successfully a year later.

A number of cantilever bridges up to 1644 ft (501 m) span have been built in America; for example:

- Commodore Barry, 1644 ft (501 m), Pennsylvania, 1974
- Greater New Orleans, 1575 ft (480 m), Louisiana, 1958
- East Bay, 1400 ft (427 m), San Francisco, 1936.

A very remarkable example of this type of construction is the Howrah Bridge in Calcutta; it had a 1500 ft (457 m) central span and 270 ft (82 m) high main towers made in steel and was completed in 1943. The Minato Bridge in Osaka, Japan, completed in 1974, has a 1673 ft (510 m) central span.

Another form of construction came to bridge the wide waterways in different parts of the world. The St Louis Bridge of Eads was the forerunner of the long-span arch type of bridge. From the later 1860s, several arch spans of up to 350 ft (107 m) were built over the Rhine in Germany. In Oporto, Portugal, two bridges, the Pia Maria and Luiz I, were built, in 1877 and 1885, respectively, the first by the famous French engineer Gustave Eiffel and the second by another Frenchman T. Seyrig. The Luiz I Bridge had a tied arch span of 560 ft (171 m); it carried a road on the top of the arch and its tie carried a rail track. Eiffel also built, in 1885, the famous Garabit viaduct in the South of France with an arch span of 540 ft (165 m) to carry a railway 400 ft (122 m) above a gorge. All these bridges had arch ribs made of wrought iron.

In Germany, the Kaiser Wilhelm Bridge at Mungsten, the Düsseldorf−Oberkassel Bridge and the Bonn−Beuel Bridge over the Rhine were built in 1897−8, of arch spans 170, 181 and 188 m (557, 595 and 616 ft),

Plate 18.   Severn Road Bridge, England.

Plate 19. Humber Bridge, England.

respectively. The first steel bridge to be built in France was the Viaur Viaduct in Southern France with a central arch span of 721 ft (220 m) carrying a railway. In 1897 the 840 ft (256 m) braced-parallel-chord arch span of the Clifton Bridge at Niagara was built, followed by the 950 ft (290 m) span box-girder arch rib of high tensile steel of the Rainbow Bridge. Another historic bridge of this form of construction deserves a mention — the railway bridge over the Zambezi river near the Victoria Falls in Africa. The 500 ft (152 m) span was built in two halves, cantilevering from each side over the 400 ft (122 m) deep gorge, by British engineers led by Sir Ralph Freeman.

The next major arch bridge was the Hell Gate Bridge in New York over the East River with a span of 977 ft (298 m). Designed by Gustav Lindenthal and completed in 1916, this was a lattice spandrel-braced two-hinged arch of high-carbon steel members and it carried four rail tracks; it is still probably the most heavily loaded (per unit length) long-span bridge in the world.

Next came the Sydney Harbour Bridge. All forms of construction for long-span bridges, namely suspension, cantilever and arch, were considered for tender competition for its construction in 1923, and the winner was the spandrel-braced two-hinge steel arch span of 1670 ft (509 m) designed by Sir Ralph Freeman and built by the Dorman Long Company of Middlesborough, England. Completed in 1932, it carried four metro-type rail tracks and a 57 ft (17 m) wide roadway with two footpaths suspended from the arch 172 ft (52 m) above water. The bridge took nearly 40 000 tons of steelwork, manufactured in England and fabricated partly in England and partly in New South Wales. Some of the steel plates and sections broke all previous records in thickness and size, and tests conducted for the material properties and strength of members provided a wealth of knowledge in steel construction. Erection was by cantilevering from each side; cranes running on the upper chord of the arch lifted up lattice members from the water to be attached to the already erected cantilever which was temporarily tied back to the banks.

At about the same time was built, what was until 1977, the longest steel arch bridge in the world – the 1675 ft (511 m) span Bayonne Bridge over the Kill Van Kull in New Jersey, designed by Othmar Ammann. The site conditions permitted the erection of this bridge by temporary trestle, i.e. cantilevering was not necessary. The present record for arch span length is held by the bridge over the New River Gorge at West Virginia, 1700 ft (518 m), built in 1977.

The great success of the suspension bridge at Brooklyn inspired the building of Williamsburg and Manhattan Bridges in New York in 1903 and 1909, the latter designed by Leon Moisseiff using the recently developed 'deflection theory for suspension bridges' by Melan and Steinman, which takes into account second-order deflections of the main cable under live load. After the First World War two more bridges of this type were built – the Camden in Philadelphia in 1926 and the Ambassador in Detroit in 1929 – reaching the span lengths of 1750 and 1850 ft (534 and 564 m), respectively. In the latter case, instead of cold-drawn wires, heat-treated wires with yield stress of 85 T/in^2 (1310 N/mm^2) (as against 64–65 T/in^2 yield stress of the former) were tried for the cables; but the discovery of broken wires where they change direction led to their replacement by cold-drawn wires.

Then came the gigantic leap of this form of construction in the shape of the George Washington Bridge over the Hudson River in New York. Designed by Othmar Ammann, its span reached 3500 ft (1067 m), nearly double the previous record, and its steel towers rose nearly 600 ft (183 m) in the air. Originally designed for a roadway of eight traffic lanes and a lower deck of railways, it was completed in 1931 without the latter and hence without the interconnecting stiffening truss. The massive weight of

Plate 20. Cable Spinning for Humber Bridge.

Plate 21. Second Bosporus Bridge, Turkey.

the deck and the cables gave it aerodynamic stability. A lower deck to carry more road traffic, and a stiffening truss, were added in 1962.

On the Pacific coast, the attraction and feasibility of bridging the sea incursions in San Francisco was exercising the minds of the bridge builders for several decades. In 1933 work commenced to bridge the Oakland Bay between San Francisco city and the mainland on the east by means of a 4 mile (6.5 km) long sea crossing of two suspension bridges each with 2310 ft (704 m) central span and 1160 ft (354 m) side spans with a common middle anchorage, a tunnel through an island, a 1400 ft (427 m) span cantilever truss bridge and approach spans, carrying eight lanes of road traffic and two metro rail tracks on double decks. Soon after, the building of the record 4200 ft (1280 m) span Golden Gate Bridge also started to connect the city with Marin County to the north across the Golden Gate Straight. Designed by J. B. Straus and completed in 1937, painted a deep red and with its 750 ft (229 m) tall portal braced towers, this is arguably the world's most scenic bridge in a spectacular setting, and its proximity to the great seismic fault made it the most daring engineering feat.

In 1940 another beautiful suspension bridge of 2800 ft (853 m) central span was opened across Tacoma Narrows in Washington State. Designed by Leon Moisseiff and carrying only two traffic lanes, the deck was 39 ft (11.9 m) wide and supported on 8 ft (2.4 m) deep plate girders rather than a lattice structure. From the opening, very substantial horizontal and vertical movements of the deck in wave forms were noticeable even in moderate wind and light traffic, and earned for the bridge a nickname 'Galloping Gertie'. Before its construction, tests in a wind tunnel had shown it to be capable of resisting gale forces of up to 120 mile/h (193 km/h). On 7 November 1940, a storm that raged for several hours and reached a speed of 42 mile/h (68 km/h) drove the bridge into an uncontrollable torsional oscillation, culminating in its collapse into the water.

After the great success of long span bridges in the previous 60 years, this disaster shook the very foundations of bridge building. The following official inquiry by three great engineering experts, von Karman, Ammann and Glen Woodruff, blamed no individuals and pointed out no mistakes; it attributed the failure to a lack of proper understanding and knowledge of the whole profession. The deck was too narrow for the span and thus its torsional rigidity was inadequate, and the plate girders not only provided insufficient flexural rigidity, but their bluff elevation caused wind vortices above and underneath the deck even in moderate and steady wind speeds.

Substantial movements in wind were previously found in the 2300 ft (701 m) span Bronx Whitestone Bridge, which had a 74 ft (23 m) wide deck, and also in the Golden Gate Bridge, and diagonal stays between the cable and the deck and additional lateral bracing in the deck structure had

Plate 22.   Knie Bridge across the Rhine, Düsseldorf, Germany.

Plate 23. Wye Bridge, Wales

to be provided. A chain pier at Brighton, England, had collapsed in a storm several years earlier.

The positive outcome of the Tacoma disaster was the extensive wind tunnel testing of scaled models and aerodynamic analysis of various deck shapes in all wind speeds. This practice re-established long-span construction on a firmer basis, leading not only to the reconstruction of the Tacoma Bridge in 1950 with a wider 60 ft (18.3 m) deck with 33 ft (10 m) deep stiffening trusses, but several more such bridges were built, e.g. Mackinac Bridge in Michigan in 1957 with 3800 ft (1159 m) span, designed by David Steinmann, and finally in 1965 the 4260 ft span (1298 m) Verrazano Narrows Bridge across the New York harbour entrance, designed by Ammann, which just exceeded the then longest span length of the Golden Gate Bridge. Steinmann introduced the concept of leaving slots in the deck, so that wind vortices escape upwards from underneath, thus setting up turbulence and thereby reducing the rhythmic up and down forces on the deck.

In Europe, Tancarville Bridge over the Seine at Le Havre with a main span of 610 m (2000 ft) was completed in 1959. The non-American features of Tancarville Bridge were the concrete towers and the continuity of the stiffening girder between the main and the side spans. This was followed in 1964 by the huge bridge over the Tagus at Lisbon with a central span of

Plate 24.   Kohlbrand Bridge, Hamburg, Germany.

Plate 25.   Sava Bridge, Belgrade, Yugoslavia.

1013 m (3323 ft) and almost at the same time the Forth Road Bridge near Edinburgh with a suspended central span of 1006 m (3300 ft). Then came the revolutionary 988 m (3240 ft) central span Severn Bridge in 1966, with its all-welded aerofoil-shaped box girder suspended structure in which the functions of a stiffening girder and a road deck were integrated, resulting in a very substantial reduction in the weight of deck steelwork and cable sizes. The hangers by which the deck is supported from the main cables were made inclined rather than vertical, thus constituting a triangulated lattice pattern; this was expected to provide additional aerodynamic damping. These concepts of the designers Freeman, Fox & Partners were repeated to bridge the Bosporus Straits by a spectacular bridge of 1074 m (3524 ft) span in 1973 and then in 1981 the record-breaking Humber Bridge in northern England with its 1410 m (4626 ft) central span. The success of the Bosporus Bridge in carrying and generating traffic has led to the building of a 1014 m (3327 ft) span second bridge which opened in June 1988 and also to a competition for a third bridge.

The great project of connecting the Japanese Honshu and Shikoku islands by road and rail bridges along three routes across the Sato Island Sea was started in the mid-1970s. One route, Kojima–Sakaide, is nearly complete; the other two are due for completion before the end of the century. This project has a number of suspension and cable-stayed bridges of giant spans, including what will be the longest suspension span of 1990 m (6530 ft) of the Akashi Kaikyo Bridge and another seven bridges with spans longer than 800 m. These bridges shall have to resist typhoons of up to 85 m/s, earthquakes of magnitude about 8 in the Richter scale, up to 100 m sea depth and 5 m/s tidal current.

In cable-stayed bridges the cables are virtually straight between their top at the tower and their bottom end at the deck where they support the deck superstructure. Thus, unlike suspension bridge cables, their tension is uniform along their length and, in this respect at least, they are more efficient. Elimination of substantial anchorages in the ground is another advantage. This type of bridge construction has become the favourite in the span range of 150–500 m, replacing suspension bridges in the higher part of this range.

Cable-stayed bridges are statically indeterminate for structural analysis; each cable stay represents one redundancy. Thus for a three span bridge, with one pair of cables supported from each tower top and two vertical cable planes, there will be eight redundancies for the eight cable supports, in addition to the two represented by the intermediate piers. Historically, several bridges were built in the first half of the 19th century, with inclined cable stays supporting the bridge span. These cables were made from bars and chains and were not initially tensioned; this allowed large deflections of the deck under loading. This shortcoming led to the concept of combining main suspension cables of a suspension bridge with a system of inclined cable stays fixed between the deck and the towers.

Arnodin in France was a pioneer of a system in which the central portion of the span was supported by suspension cables, but the end portions near the towers were held by cable stays radiating from the towers. The Franz Joseph Bridge in Prague (1868), the Albert Bridge over the Thames in London (1873), the Ohio River Bridge at Cincinnati (1867), and the Niagara (1855) and Brooklyn (1883) Bridges by Roebling were examples of the concept of combined suspension and cable stay system. The cable stays not only took a substantial portion of the vertical dead and live loading, but also provided the crucial aerodynamic stability. The Lezardrieux Bridge over the Trieaux River in France built in 1925 is the first known example of the modern elegant cable-stay system, where the cable radiated from the tower tops and transferred their tension to the stiffening girders.

After the Second World War, the need for the reconstruction of the war-damaged bridges in Europe while building materials were in short supply led the designers to this form of construction. In Germany, Dis-

Plate 26.   Auckland Harbour Bridge, New Zealand.

Plate 27.　Milford Haven Bridge during construction.

chinger carried out extensive studies and concluded that cables formed with high strength wires and substantially pre-tensioned to support the dead load of the deck would provide adequate stiffness and aerodynamic stability; it is also essential to achieve accurate tensioning of the cable along with the desired profiles of the spans under their dead loads.

Dischinger designed and German engineers built the first bridge of this kind, the Strömsund Bridge in Sweden, opened in 1956, with three spans of 75−183−75 m (246−600−246 ft) and two cable stays radiating from each tower top in each direction in a fan arrangement along a vertical plane near each edge of the bridge deck. The stiffening girder consisted of two plate girders along the cable planes; with this form of deck construction, the transverse distribution of deck loading between the two cable planes could be based on statics only; this reduced the cable redundancies by half and was just about manageable for calculation by slide rules. The width of the navigation channel along the river Rhine often demanded clear spans of over 250 m (820 ft) even during erection, and this new bridge type made this economically possible.

The Theodor Heuss Bridge across the Rhine at Düsseldorf, opened in

1957, had spans of 108−260−108 m (354−853−354 ft) and three sets of parallel cables from each tower in each direction, supported from three points in the tower height in what is now called a harp arrangement. The stiffening girder consisted of two box girders along the cable planes; their torsional rigidity influenced the transverse distribution of deck loading between the cable planes, thus doubling the structural redundancies. An orthotropic steel deck spanned between the longitudinal girders. This bridge set in motion an impressive variety of cable-stay bridge construction in post-war Germany.

The next bridge, the Severins across the Rhine in Koln, opened in 1960 and became famous for its single A-shaped tower on one bank of the Rhine and two unequal spans of 302 and 151 m (991 and 495 ft); it had three pairs of cables on each side of the tower arranged in a fan shape along inclined cable planes.

The third German bridge, across the Elbe River in Hamburg, introduced the concept, in 1962, of a single cable plane with a central torsionally stiff stiffening girder of box type along the longitudinal axis of the bridge,

Plate 28. Avonmouth Bridge, England.

though it also had an outer longitudinal plate girder at each edge of the deck. The peculiarities of this bridge were the upward extension of the two towers to double their height above the top cable connection purely for appearance, and the reverse fan arrangement of the two cables on each side of the two towers anchored to a common attachment point at the deck but connected at two different heights of the tower.

Then came the classical Leverkusen Bridge across the Rhine in 1964, with a central cable plane and two cables on each side of two towers in a harp arrangement to support three 106−280−106 m (348−919−348 ft) spans.

In the late 1960s the introduction of computers for the analysis of redundant structural systems heralded the multi-cable system of stays, whereby a large number of small cables attached to the towers at various heights in fan or harp or a combined fashion support the bridge deck at close intervals. This evolution simplified the construction of each cable and its end connections, reduced the stiffness requirement of the stiffening girder, which became virtually a beam on continuous elastic supports, and thus increased the span range of this form of bridge construction.

Plate 29.   Europa Bridge, Italy.

The first of the multi-cable bridges was the Friedrich Ebert Bridge across the Rhine at Bonn, completed in 1967, with a single cable plane containing 80 cables, supporting a wide box stiffening girder over 120−280−120 m (394−919−394 ft) spans, followed closely by the Rhine Bridge at Rees, with two cable planes and two plate girders as the stiffening girder.

In the Knie Bridge across the Rhine at Düsseldorf, opened in 1969, cables in the side spans were anchored to the piers underneath; by increasing the longitudinal rigidity of the whole structure, this innovation enabled the construction of a 320 m (1050 ft) long span over the river supported by cables from only one tower; if supported from two towers, the span could conceivably be doubled! The same technique was used to build the symmetrical 350 m (1148 ft) span Duisburg−Neuenkamp bridge over the Rhine in 1970.

The Erskine Bridge in Scotland, opened in 1971, had a large 305 m (1000 ft) long span but, following the Wye Bridge design of the early 1960s, employed only one cable on either side of the two towers along a central vertical plane. The 325 m (1066 ft) span Kohlbrand in Hamburg is the first bridge with multiple cables arranged in inclined planes from A-shaped towers. Other remarkable cable-stayed steel-deck bridges are:

(1) over the Waal near Ewijik, Holland, 270 m (886 ft) span completed in 1975
(2) Düsseldorf Flehe bridge over the Rhine at Düsseldorf, Germany, 367 m (1204 ft), 1978
(3) Stretto di Rande at Vigo, Spain, 400 m (1312 ft), 1978.

The first double-decked cable-stayed bridge was built in 1977 in Japan; the Rokko Bridge had a truss stiffening girder of 8 m (26 ft) depth to provide the necessary height and light on the lower deck. The first bridge with cable stays anchored to the ground was the Indiano Bridge over the Arno river in Florence. The first cable-stayed bridge to support a rail track was the (twin) bridge(s) across the Parana River in Argentina built in 1978, followed by the bridge over the Sava River in Belgrade with a main span of 254 m (833 ft) carrying two heavy railway tracks.

The Tjörn Bridge in Sweden, completed in 1982, has a 366 m (1201 ft) main span high above water; in fact this bridge was built to replace a steel arch bridge of 280 m (918 ft) which was demolished in a collision with a ship at a low point on the arch. The St Nazair Bridge completed in 1975 in Brittany, France, has a central span of 404 m (1325 ft), and Faro Bridge in Denmark, completed in 1985, has a 295 m (968 ft) span. The Luling Bridge across the Mississippi near New Orleans, opened in 1984, has a 25 m (82 ft) wide steel orthotropic deck on a trapezoidal box girder, supported by 12 cable stays from each A-frame tower, with a 376 m (1235 ft) central span. The Meiko Nishi Bridge in Japan, completed in 1985, has a 405 m (1329 ft) span.

The Annacis Bridge in Vancouver, Canada, and the Dao Kanong Bridge in Bangkok, Thailand, were opened in 1987 and had central spans of 465 and 450 m (1526 and 1476 ft), respectively. Currently under construction are the Tampico Bridge in Mexico with 360 m (1181 ft) span, the Hooghly Bridge in Calcutta, India, with a 457 m (1500 ft) span, the Yokohama Bay Bridge in Japan with 460 m (1509 ft) span to carry 12 lanes of traffic on two decks, and work has also started for a 450 m (1476 ft) span bridge over the Thames at Dartford, near London.

The Honshu−Shikoku Bridge project has a fascinating pair of cable-stayed bridges located end-to-end, each with 185−420−185 m (607−1378−607 ft) spans to carry roadway on the upper deck and railway on the lower; there is also another bridge called Ikuchi with 150−490−150 m (492−1608−492 ft) cable-stayed spans. A second bridge 5 km downstream of the Severn Suspension Bridge between England and Wales is proposed to have a 450−460 m (1475−1510 ft) cable-stayed span. Finally, a cable-stayed span of 816 m (2677 ft) has been proposed to bridge the Seine at

Plate 30. Foyle Bridge, Northern Ireland.

Plate 31.   A motorway bridge at Brentwood, England.

Normandy, France. For cable-stayed bridges with concrete deck girders, mention should be made of the Brotonne Bridge at Rouen in France opened in 1977 with 320 m (1050 ft) span, Sunshine Skyway at Tampa Bay, Florida, opened in 1987 with 366 m (1200 ft) span, and Barrios du Luna in Spain with 440 m (1444 ft) span.

In the early days of steel bridge construction, riveting and bolting were the means of connecting component parts in plate and trussed girders. The box girder type of construction was the exception to the general practice, e.g. Stephenson's railway bridges at Menai and Conway. Electric open-arc welding was developed in the 1930s and the Second World War saw its rapid expansion. Then followed the trend of stiffened plate construction and friction-grip bolting. Rivets gradually gave way to shop welding and friction-grip bolting on site. Erection methods of launching and cantilevering were widened by the development of floating-out techniques and heavy lifting equipment suitable for handling entire structures in one piece.

Post-war years also saw a great expansion of the understanding of structural behaviour and analysis of indeterminate interconnected structural systems. The state-of-the-art before the war generally consisted of assuming pin-jointed connections between different structural elements and manually solving simultaneous equations. In the aftermath of the war devastations, the knowledge and experience of the aircraft industry was imported into bridge building in order to provide the expertise necessary for rebuilding thousands of demolished bridges with minimum amounts of scarce construction materials like steel. This brought in the understanding of torsional behaviour of thin-walled closed sections, and many of the mathematical tools for solving complex analytical problems. This was followed by the advent of electronic computers which vastly increased the facilities for structural analysis. Highly indeterminate structural systems could now be analysed in minutes and it was no longer necessary to make assumptions like pinned joints. One immediate benefit was the improved lateral distribution of concentrated loads over the whole width of the bridge deck. Orthotropic steel decks became the favourite type of light-weight economical bridge construction over 500 ft (150 m) spans. The torsional stiffness of single or multiple cell box girders proved highly advantageous in the 'cantilever' type erection method over great heights or wide rivers. The Düsseldorf−Neuss Bridge across the Rhine, with 103−206−103 m (338−676−338 ft) span box girders with a steel orthotropic deck and completed in 1951, set this modern trend. The Sava Road Bridge in Belgrade, Yugoslavia, with continuous plate girders of 856 ft (261 m) centre span completed in 1956 was another landmark. In box girders with orthotropic top flange, each element functioned in multiple ways, e.g. the flange stiffeners carried the wheel loads and also acted as the tension/compression flange of the box girder; this led to great economy of material. Great examples of this form of construction, with their maximum span length and year of completion, are:

- Zoo Bridge, Cologne, West Germany, 850 ft (259 m) 1966
- Charlotte Bridge, Luxembourg, 768 ft (234 m) 1966

Plate 32. Windmill Bridge, Newark, England.

Plate 33.   Rio Niteroi Bridge, Brazil.

- San Mateo Hayward, San Francisco, 750 ft (229 m) 1967
- Auckland Harbour, New Zealand, 800 ft (244 m) 1969
- Gazelle, Belgrade, Yugoslavia, 820 ft (250 m) 1970.

The 250 m (820 ft) high, 376 m (1234 ft) span motorway bridge over the Sfalasse Valley in Calabria, Italy, with two inclined portal legs that divided the span into 110−156−110 m (361−512−361 ft), completed in the early 1970s, was another striking example of steel box girder orthotropic deck construction.

However, the systematic and confident progress with light and rapid construction of steel bridges suffered a big jolt in the early 1970s, with the failure of three big box girder bridges in Britain, Australia and Germany. With hindsight it is clear that sufficient attention was not being given to some of the details of box girder design and construction, particularly during construction when the box girder was not completely built. The research that followed these failures clarified the buckling behaviour of stiffened plates under complex stress patterns of combinations of compression, shear and bending and the effects of unavoidable initial geometrical out-of-flatness and out-of-straightness and of welding residual stresses.

Standards and codes for steel bridge design and construction were updated to take advantage of these developments; BS 5400 in Britain took the lead in these advances in the late 1970s and early 1980s. As a record-span steel box girder bridge, mention should be made of the Costa e Silva (or Rio−Niteroi) Bridge across the Guanabara Bay in Brazil with 200−300−200 m (656−984−656 ft) spans designed by American designers and built by a British consortium in 1973.

Chapter 2
Types and Properties of Steel

2.1 Introduction

Steel used for building bridges and structures contains:

(1) iron
(2) a small percentage of carbon and manganese
(3) impurities that cannot be fully removed from the ore, namely sulphur and phosphorus
(4) some alloying elements that are added in very small quantities to improve the properties of the finished product, namely copper, silicon, nickel, chromium, molybdenum, vanadium, columbium and zirconium.

The strength of the steel increases as the carbon content increases, but some other properties like ductility and weldability decreases. Sulphur and phosphorus have undesirable effects and hence their maximum amount is controlled. Steel used for building bridges may be grouped into the following three categories:

(1) Carbon steels — Only manganese, and sometimes a trace of copper and silicon, are used as alloying elements. This is the cheapest steel available for structural uses where rigidity rather than strength is important. It comes with yield stress up to 275 N/mm^2 and can be easily welded. The American ASTM A36, the British Grades 40 and 43, and Euronorm 25 Grades 235 and 275 steels belong to this category.

(2) High strength steels — These cover steels of a wide variety with yield stress in the range of 300 to 390 N/mm^2. They derive their higher strength and other required properties from the addition of alloying elements mentioned earlier. British Grade 50, American ASTM A572 and Euronorm 155 Grade 360 steels belong to this category. Another variety of steel in this category is produced with enhanced resistance to atmospheric corrosion

42

and these can be left unpainted in appropriate situations. These are called 'weathering' steel in Europe; in America they come under ASTM A588 and have various trade names like 'Cor-ten'.

(3) Heat-treated carbon steels — These are the steels with the highest strength, and still retain all the other properties that are essential for building bridges. They derive their enhanced strength from some form of heat treatment after rolling, namely normalisation or quenching-and-tempering.

2.2 Properties

The properties of structural steel relevant for its use in bridge construction are the following:

(1) strength
(2) ductility
(3) weldability
(4) notch toughness
(5) weather resistance.

The strength properties of commonly available structural steels are represented in the idealised tensile stress−strain behaviour in Fig. 2.1(a). The slope of the initial linear part is defined as Young's modulus E. At a stress just beyond the limit of linearity, the flow of the steel becomes plastic at nearly constant stress. This stress is called the yield stress (or yield point) R_e of the steel. After the yield is completed the stress increases again until the maximum stress, called the tensile strength (R_m) is reached. With further straining large local elongation and reduction in cross-section occur, and the stress falls until fracture takes place.

With some steel, after the initial limit of linearity, the stress may attain a maximum, then fall and remain approximately constant during yielding, as shown in Fig. 2.1(b); the value of the stress at the commencement of yield is called the upper yield stress R_{eH}. Some steel does not show the yield phenomenon; beyond the limit of linearity, the strain continues to increase non-proportionally, as shown in Fig. 2.1(c). In such cases a 'proof stress' is measured. Proof stress (total elongation) R_t is measured by drawing a line parallel to the stress axis and distant from it by the required total elongation; proof stress (non-proportional elongation) R_p is measured by drawing a line parallel to the initial straight portion of the behaviour and distant from it by the required non-proportional elongation. For such steel a 0.5% total elongation proof stress is regarded as the yield stress.

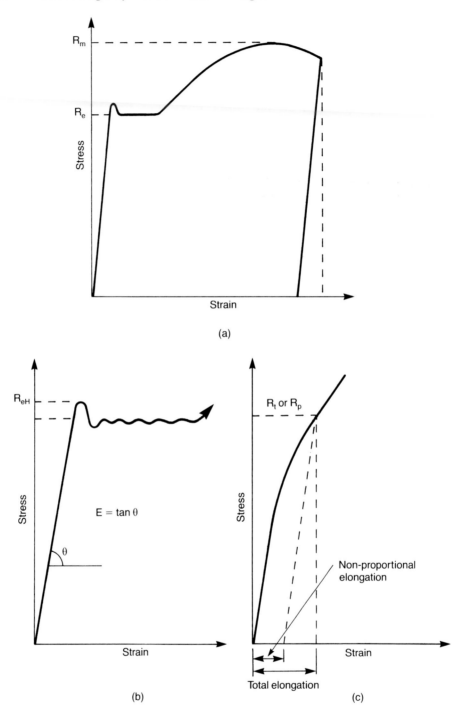

Fig. 2.1 Idealised stress−strain behaviour of steel.

Unloading from any stage of initial straining occurs along a line approximately parallel to the initial straight portion of the stress−strain curve.

2.3 Yield stress

The yield stress is the most important strength parameter of structural steel. Yield stress is normally measured in tension only. Such measurements are affected by the specimen geometry, the rate of straining, location and orientation of the specimen in the rolled section or plate and the stiffness of the testing machine. A higher rate of straining increases the yield stress and the tensile strength. Various national and international standards specify these testing parameters; for example, for the determination of yield stress British Standard BS 18[1] limits the rate of straining at the time of yielding to 0.0025 per second; when this cannot be achieved by direct control, the initial elastic stressing rate has to be controlled within the values stipulated for different testing machine stiffnesses.

Within a zone of, say, 50 mm from a rolled edge, yield stress may be up to 15% higher than in the remainder of a plate. Yield stress in the transverse direction may be approximately $2\frac{1}{2}\%$ less than in the longitudinal direction of rolling.

The yield stress (and also the tensile strength) varies with the chemical composition of the steel, the amount of mechanical working that the steel undergoes during the rolling process and the heat treatment and/or cold working applied after rolling. Thinner sections produced by an increased amount of rolling have higher yield stresses; even in one cross-section of a rolled section the thinner parts have higher yield stresses than the thicker parts. Heat treatment or cold working may remove the yield phenomenon.

The stress−strain behaviour under compression is normally not determined by tests and is assumed to be identical to the tensile behaviour. In reality the compressive yield stress may be approximately 5% higher than the tensile yield stress. The state of stress at any point in a structural member may be a combination of normal stresses in orthogonal directions plus shear stresses in these planes. Several classical theories for yielding in three-dimensional stress states have been postulated; the theory that has been found most suitable for ductile material with similar strength in compression and tension is based on the maximum distortion energy and attributed variously to Huber, von Mises and Hencky. According to this theory, in a two-dimensional stress state yielding takes place when normal stresses σ_1 and σ_2 on the two orthogonal planes and shear stress τ on these planes satisfy the following condition:

$$\sigma_1^2 + \sigma_2^2 - \sigma_1\sigma_2 + 3\tau^2 = \sigma_y^2$$

where σ_y is the measured yield stress of the material. It may be noted that, according to this theory: (i) the yield stress τ_y in pure shear, i.e. without any normal stresses, is equal to $\sigma_y/\sqrt{3}$ and (ii) in the biaxial stress state, the normal stress in one direction may reach values higher than the measured uniaxial yield stress σ_y before yielding takes place; e.g. if $\sigma_1 = 2\sigma_2$, yielding will not take place till σ_1 reaches approximately 15% higher than the uniaxial yield stress of the material.

The other elastic properties that influence the state of stress at any point are:

(1) Young's Modulus E, which is in the range 200 to 210 kN/mm^2
(2) The shear modulus or modulus of rigidity G, which is the ratio between shearing stress and shear strain and is in the range 77 to 80 kN/mm^2
(3) Poisson's ratio μ, which is the ratio between lateral strain and longitudinal strain caused by a longitudinally applied stress and is usually taken as 0.3 for structural steel
(4) The coefficient of thermal expansion which is the expansion/contraction per unit length caused by one degree change in the temperature and is normally taken as $12 \times 10^{-6}/°C$.

There is a theoretical relationship between E, G, and μ, given by

$$G = \frac{E}{2(1 + \mu)}$$

2.4 Ductility

Ductility of a material is measured by its capacity to undergo large strains after the onset of yielding and before fracture. This property enables a structure to exhibit large deformation when the load it carries exceeds the value corresponding to its yield stress, thus providing an advance warning of possible failure.

In a structure with redundancy, i.e. alternative load paths, when yield strain is exceeded in a critical component, this property enables it to redistribute the excess load to other components while the critical component retains its yield load. In a tensile test, after the maximum load, i.e. tensile strength R_m (see Fig. 2.1(a)), is reached, necking occurs in one cross-section, accompanied by large local elongation.

After fracture, the two pieces are held together to measure the total permanent elongation of the original gauge length. This elongation represents the ductility of the material. However, it is affected by the geometrical parameters of a test piece, e.g. cross-sectional shape and gauge length. By international agreement, the relationship between gauge length L_o and cross-sectional area S_o in the gauge length of tensile test pieces has been established as $L_o = 5.65\sqrt{S_o}$. However, other gauge lengths are often used, e.g. 200 mm or 8 inches are quite common in the UK and USA, respectively. Structural steel specifications prescribe a minimum value of this percentage elongation, in the range of 18 to 25%.

   Good ductility through the thickness of thick rolled products is necessary to prevent lamellar tearing when high stresses occur in the direction of the thickness.

2.5   Notch ductility

Notch ductility is a property of metals indicating their resistance to brittle fracture. Brittle fracture is a form of failure that occurs suddenly under a load well below the level to cause yielding. It is initiated by the existence of a small crack or other form of notch. Very high concentration of stress occurs at the root of a natural crack. Any sudden change in the cross-section of a loaded member has a notch-like effect, namely it disturbs the stress pattern and causes a local stress concentration. If the local yielding at the tip of the crack or notch is insufficient to spread the load over a large area, a brittle fracture may be initiated. Once initiated, the fracture propagates at high speed driven by the release of the elastic strain energy in the structure.

   Design should avoid any sharp geometrical discontinuity, change of section and re-entrant angles. Workmanship should avoid accidental notches like dents, pitting and various weld defects, namely undercutting, slag inclusions, porosity or cracks.

   Brittle fracture is more likely to occur at low temperatures, as the notch ductility of steel falls with decreasing temperature. Sometimes a temperature change of a few degrees changes the ductility so substantially that there is a transition from a ductile to brittle type behaviour.

   Brittle fracture is more likely in a massive structural component than a light one. This is due to the three-dimensional stress conditions in thick elements and the likelihood of non-metallic inclusions, segregation or lamination left in thick rolled products.

   Welded steelwork has more propensity to brittle fracture than unwelded steelwork, as welding may:

(1)   introduce defects
(2)   reduce the notch ductility of the heat-affected zone near the weld
(3)   introduce weld metal of different notch ductility
(4)   leave substantial residual stresses, particularly tensile stresses as high as yield stress near the welds.

Flame cutting also produces defects and reduces ductility in the hardened heat-affected zone.

Other factors that increase susceptibility to brittle fracture are:

(1) cold working during the fabrication process, namely bending, shearing, etc., particularly in higher strength steels
(2) 'hot dip' galvanising
(3) impact loading
(4) unsuitable heat treatment
(5) large amount of non-metallic alloying elements in the steel.

There is no single measurable property of structural steel by which its susceptibility to brittle fracture can be uniquely measured. The test that is commonly used nowadays is called the Charpy V-notch impact test. In this test a specimen is hit by a striker mounted on the end of a pendulum. The striker is lifted initially to a specified height and then released to hit and break the specimen, swinging and rising to a height on the other side. The difference between the two heights, multiplied by the mass of the striker, is the energy absorbed in fracturing the specimen. The specimen has a notch at the point of maximum tensile stress, and is held at a specified temperature.

The theory of fracture mechanics seeks to explain the static strength of a component in terms of the size of a pre-existing notch or flow. This theory was originally restricted to elastic stress conditions in thin sheets, i.e. without any stress in the through-thickness direction, but has been extended more recently to thick plates in which the strain, but not the stress, through the thickness is negligible, and also to allow for yielding or plastic deformation before fracture. The fracture mechanics theory attempts to establish a relationship between the fracture toughness of the material, the permitted size of the initial crack or flaw and the permitted stress level.

In British Standard BS 5400 Part 3[2] for steel bridges, the minimum notch ductility requirements for welded construction with substantial residual tensile stress has been specified as

$$C_v = \frac{\sigma_y}{355} \frac{t}{2}$$

where C_v is the energy absorption in joules in the Charpy V-notch test at the minimum service temperature of the structure, σ_y is the yield stress in N/mm^2 and t the thickness in millimetres of the component. In this relationship it has been assumed that welding produces residual tensile stress of the magnitude of the material yield stress and that the applied stress level due to service load is two-thirds of the yield stress. A higher energy absorption is specified for situations of high stress concentration, where the plastic strain on reaching yield may be several times the yield strain. The minimum temperature of a steel bridge structure is usually several degrees centigrade lower than the minimum shade air temperature of the site.

In most countries, structural steel for each level of yield stress is produced in several grades of Charpy energy absorption. These grades are either in an increasing order of energy value for the same test temperature or constant energy value for decreasing levels of test temperatures or a mixture of both. Unfortunately no unique relationship exists between energy values at different test temperatures. British Standard BS 4360[3] for structural steel provides several alternative sets of test requirements for energy and temperatures for the same grade of steel. From this general pattern and the energy requirement at service temperature given above, limiting thicknesses of various grades of steel to BS 4360[3] for various minimum service temperatures are tabulated in BS 5400 Part 3[2].

The Charpy V-notch impact requirements in the AASHTO specification are 15 ft lb or 27 joules at a test temperature 70°F (about 39°C) higher than the minimum service temperature of the bridge. This shift in the temperature is to reflect the fact that the Charpy test is conducted at a rapid rate of loading, whereas the actual loading rate in a bridge component is much slower. Three minimum service temperature zones are given, with 0°F and above (−17.8°C and above) for zone 1, −1° to −30°F (−18.3° to −34°C) for zone 2 and −31° to −60°F (−35° to −51°C) for zone 3. It is understood that these impact requirements are now the subject of a review and more stringent values may be specified in future.

2.6   Weldability

Steels with carbon content less than about 0.30% are weldable, provided suitable welding processes and electrodes are used. This includes all the categories of steel used for building bridges and mentioned in Section 2.1. However, even for these steels, ease of weldability varies. Increased amounts of carbon and manganese, which are necessary for higher strengths, make the steel harder and consequently more difficult to weld. The group of alloying elements, chromium, molybdenum and vanadium, which are added to further increase the strength, and the elements nickel and copper which are added to improve weathering resistance, also reduce weldability. For the purpose of measuring weldability, a term 'carbon equivalent' is used, which is given by the following formula:

$$C + \frac{Mn}{6} + \frac{Cr + Mo + V}{5} + \frac{Ni + Cu}{15}$$

where C, Mn, etc. represent the percentage of the element concerned in the chemical composition of the steel. The chemical composition of steel, i.e. the percentage content of the various alloying elements, is usually checked by ladle analysis. Chemical analysis may also be done on the finished product itself, but some deviations as stipulated in the specifications are then permitted from the specified composition limits. Higher yield stresses can only be obtained by increasing the percentage content of the various alloying elements, thus increasing the carbon equivalent value. Hence welding of higher strength steels is more difficult. Maximum values for carbon equivalent are sometimes given in the specifications. Steels with carbon equivalent values higher than 0.53 are likely to require special measures in welding. Welding difficulties also increase with thicker members and increased restraint against shrinkage due to cooling. If the cooling rate is too rapid, excessive hardening and cracking may occur in the weld and the heat affected zone: this can be avoided by controlling weld run dimensions in relation to material thickness, by applying pre-heat and controlling interpass temperature between weld runs and by using low-hydrogen electrodes.

2.7   Weather resistance

A special type of steel has been developed for increased resistance to corrosion — this is called 'weathering steel'. When used in the appropriate environment, this steel forms a thin iron oxide film on the surface. This film is a tightly adherent coating that resists any further rusting by preventing the ingress of moisture. Alternate drying and wetting is the ideal condition for the formation and durability of this film. Under bridges in countries like Britain, the environment is usually damp most of the time, and this film comes off and is reformed at intervals of one to three years. Weathering steel bridges can be left unpainted if the colour of the oxide coating is aesthetically acceptable — it varies from red—orange to purple—brown. In Britain, the Department of Transport specifies[4] an additional sacrificial thickness of one or two millimetres on each surface, depending on the sulphur content in the atmosphere; unpainted weathering steel is not permitted: (i) in a marine environment, (ii) where it can be subjected to de-icing salt spray and (iii) in locations with very high atmospheric sulphur content. In the United States, weathering steel is known by trade names like Mayri R (Bethlehem Steel) and Cor-ten (United States Steel).

Higher corrosion resistance of weathering steel is achieved by adding chromium, copper and sometimes nickel as alloying elements; as a result the carbon equivalent value is higher, thus requiring special measures in welding.

2.8 Commercially available steels

The material properties of some commercially available steel to various national/international specifications and suitable for use in steelwork for bridge structures are tabulated in Table 2.1. The following notes apply to their properties given in this Table:

(1) BS 4360 steel:
(a) In each grade of steel there are several subgrades A to E with an increasing order of notch toughness; alternative notch toughness at different testing temperatures are also available.
(b) Yield stress varies within the range shown with thickness and subgrades.
(c) Maximum carbon equivalent values are available for subgrade C and above, and only if specified.
(2) Steel to ASTM A36, A572 and A588:
(a) Only those steels are included in the Table that are covered in the AASHTO[10] Specification for highway bridges. Steel to ASTM A242 is similar to ASTM A588; steel to A441 is similar to A588 in strength but intermediate between A36 and A588 in weathering resistance; ASTM A572 covers several grades of steel with yield stress ranging from 290 to 450 N/mm^2; ASTM A514 and A517 are for high yield quenched and tempered steel with yield stress 620–690 N/mm^2; ASTM A709 covers several grades of steel similar to A36, A572 grade 50, A588 and A514 steel; they are also available in weathering quality.

Table 2.1 Properties of some commercially available steel.

Specification	Grade	Yield stress N/mm^2	Minimum strength N/mm^2	Charpy toughness	% elongation on gauge length 200 mm	% elongation on gauge length 5.65$\sqrt{S_o}$	Maxm. carbon eqv. %
BS 4360[3]	40	185–260	400	up to 27 J at −50°C	22	25	0.39 to 0.41
	43	225–275	430	up to 27 J at −50°C	20	22	0.39 to 0.41
	50	305–390	490	up to 27 J at −60°C	18	20	0.39 to 0.47
	55	400–450	550	27 J at 0 to −60°C	17	19	0.41 to 0.51
	WR50 (weathering)	325–345	480	up to 27 J at −15°C	19	21	0.54
ASTM A36[5]	−	250	400	As per supplementary requirements	20	−	Not available
ASTM A572[6]	50	345	450		18	−	
ASTM A588[7] (weathering)	−	290–345	435–485		18	−	
Euronorm 25[8]	235	185–235	340	27 J at 20° to −20°C	−	22–26	Not available
	275	225–275	410	27 J at 20° to −20°C	−	18–22	
	355	305–355	490	27–40 J at −20°C	−	18–22	
Euronorm 155[9] (weathering)	360	195–235	360	27 J at 20° to −20°C	−	24–26	Not available
	510	315–355	510	27 J at 20° to −20°C	−	20–22	

    (b) Yield stress and tensile strength of A588 steel reduce with thicker steel, within the range shown.

(3) Steel to Euronorm 25 and 155:

    (a) Yield stress and percentage elongation reduce with thicker steel.

    (b) In each grade there are several subgrades with different values of notch toughness within the range shown.

    (c) For Euronorm 25 only weldable grades of steel are included in this table.

2.9   Recent developments

The off-shore structures in the North Sea have given rise to higher requirements on structural steel than any other previous applications − due to the great water depths, huge wave forces and low temperatures encountered there. High-strength carbon-manganese steel has been improved to obtain yield stresses up to 500 N/mm^2 by the addition of very small quantities of alloying elements like niobium, vanadium, nickel, copper and molybdenum.

Quenching with pressurised water and tempering have been used to further improve the mechanical properties, with minimum yield stresses in the region of 700 N/mm^2 and good ductility, weldability and notch toughness properties.

2.10   References

1. British Standard BS 18: 1971: Methods for Tensile Testing of Metal: Part 2: Steel (General). British Standards Institution, London.
2. British Standard BS 5400: Part 3: 1982: Code of Practice for Design of Steel Bridges. British Standards Institution, London.
3. British Standard BS 4360: 1986: Specification for Weldable Structural Steel. British Standards Institution, London.
4. Departmental Standard BD7/81: Weathering Steel for Highway Structures. Department of Transport, London.
5. ASTM Designation A36−81[a]: Standard Specification for Structural Steel, January 1982. ASTM, 1916 Race Street, Philadelphia, PA 19103, USA.
6. ASTM Designation A572−82: Standard Specification for Low-Alloy Columbium-Vanadium Steels of Structural Quality. July 1982. ASTM, 1916 Race Street, Philadelphia, PA 19103, USA.
7. ASTM Designation A588−82: Standard Specification for High-Strength Low-Alloy Structural Steel with 50 Ksi Minimum Yield Point to 4 inches Thick. September 1982. ASTM, 1916 Race Street, PA 19103, USA.
8. Euronorm 25. 1972. Hot Rolled Products for General Structural Applications in Unalloyed Steel. Office for Official Publications of the European Communities, L-2985 Luxemburg.
9. Euronorm 155−80. 1980. Weathering Steel for Structural Purposes − Quality Standard. Office for Official Publications of the European Communities, L-2985 Luxemburg.
10. Standard Specification for Highway Bridges, 12th edn. (1977). American Association of State Highway and Transportation Officials.

Chapter 3
Loads on Bridges

3.1 Dead loads

The dead load on a bridge consists of the weight of all its structural parts and all the fixtures and services like deck surfacing, kerbs, parapets, lighting and signing devices, gas and water mains, electricity and telephone cables. The weight of the structural parts has to be guessed at the first instance and subsequently confirmed after the structural design is complete. The unit weights of the commonly used materials are given in Table 3.1.

3.2 Live loads

Bridge design standards of different countries specify the design loads which are meant to reflect the worst loading that can be caused on the bridge by traffic permitted and expected to pass over it. The relationship

Table 3.1 Unit weights of commonly used bridge materials.

Material	Unit weight(kg/m^3)
Steel or cast steel	7850 (77 kN/m^3)
Aluminium	2750
Cast-iron	7200
Wrought iron	7700
Reinforced concrete	2400
Plain concrete	2300
Bricks	2100–2400
Stone masonry	2200–2950
Timber	480–1200
Asphalt	2300
Tarmacadam	2400
Compacted sand, earth or gravel	1950
Loose sand, earth or gravel	1600

between bridge design loads and the regulations governing the weights and sizes of vehicles is thus obvious, but other factors like traffic volume and mixture of heavy and light vehicles are also relevant. Short spans, say up to 10 m for bending moment and 6 m for shear force, are governed by single axles or bogies with closely spaced multiple axles. The worst loading for spans over about 20 m is often caused by more than three vehicles.

The worst vehicles are often the medium weight compact vehicles with two axles, and not the heaviest vehicles with four, five or six axles. The criteria thus change from axle loads to worst vehicles as the span increases, with the mixture of vehicles in the traffic being an important factor for the longer spans. When axles or single vehicles are the worst case, the effect of impact has to be allowed for, but several closely spaced vehicles represent a jam situation without significant impact. The adjacent lanes of short span bridges may all be loaded simultaneously with the worst axles or vehicles, but this is less likely for long spans. Apart from the design loading for normal traffic, many countries specify special bridge design loading for the passage of abnormal vehicles of the military type or carrying heavy indivisible industrial equipment like generators. The passage of such heavy vehicles on public roads usually involves special permits from the highway authorities and often supervision by the police. In addition to these legal heavy loads, there is the growing problem of illegal overweight vehicles weighing as much as 40% over their legal limits.

In each country traffic regulations limit the wheel and axle loads and gross vehicle weights, and impose dimensional limits on axle spacing and size of vehicles. Goods vehicles may be of the following types:

- vehicles with two axles
- rigid vehicles with three or more axles
- articulated vehicles with two or three axles under the tractor and one or more axles under the trailer
- road trains comprising a vehicle and trailer.

The maximum legal axle weights[1] are 9.1 tonnes in the USA, 10.0 tonnes in Japan, W. Germany, Scandinavia, Holland, Canada and Switzerland, 10.5 tonnes in the United Kingdom (increased to this value in 1983), 12.0 tonnes in Italy and 13.0 tonnes in Belgium, France and Spain. The total weight limit for tandem axles is 15.4 tonnes in the USA, 16.0 tonnes in Scandinavia, Germany and Holland, 18.0 tonnes in Switzerland, 19.0 tonnes in Italy, 19.8 tonnes in Canada, 20.0 tonnes in Belgium, 20.3 tonnes in the UK, and 21.0 tonnes in France and Spain. The gross vehicle weight limits in these countries vary from 12.7 tonnes to 20 tonnes for two-axle vehicles, 19.1 tonnes to 28.1 tonnes for three-axle vehicles, 21.8 tonnes to 50.0 tonnes for articulated vehicles and 32.5 tonnes to 63.5 tonnes for road trains.

Maximum widths, lengths and heights are also controlled by these regulations, but for bridge loading minimum possible lengths and widths are the critical parameters. For example, in Britain the four-axle 30.5 tonne 8.3 m long rigid vehicle is more critical on bridge spans of 10 to 40 m than the five-axle 38 tonne articulated lorry, which must be at least 12 m long. There are currently proposals in the European Community to standardise the weight limits of goods vehicles in the ten countries. It can be seen from the above figures that at present the weight limits on vehicles are very diverse in different countries.

3.3 Design live loads in different countries

Bridge design loadings in different countries vary a great deal and they do not necessarily follow the patterns of the vehicle weight limits in the countries.

In the United Kingdom, bridge design loading is given in BS 5400 Part 2: 1978[2], with some amendments to it prescribed by the government authority. Type HA loading consists of a uniformly distributed lane loading, together with a knife edge loading of 120 kN placed across the lane width. For loaded lengths in the direction of the traffic up to 30 m, the value of the uniformly distributed lane loading is 30 kN/m; for greater loaded length (L) it is given by $151(L)^{-0.475}$, but not less than 9 kN/m. As an alternative to HA loading for short loaded lengths and areas, a wheel load of 100 kN distributed over a circular or square area with pressure of 1.1 N/mm^2 has to be considered. An impact factor of 1.25 on any one axle of one vehicle has been taken into account in the prescribed design loading. The whole carriageway, including hard shoulders if any, is divided into an integral number of notional traffic lanes of width not less than 2.3 m or more than 3.8 m. Any two such lanes are to be loaded with the full HA loading, the remaining lanes with one-third HA loading, which was increased to 0.6 HA by the government in 1984[3]. Type HB loading represents abnormal vehicles; it is a hypothetical vehicle with 16 wheels on four axles; the heaviest HB loading is for 45 units representing 180 tonnes.

HA loading was originally derived in the 1950s for the drafting of the first British Bridge Code BS 153[4]. It was based on the following sets of vehicles:

Vehicle sets for design loading in BS 153.

Loaded length		Vehicles	Gap[1] (ft)	Total weight (tonnes)	UDL (tonnes/m)
ft	m				
75	23	3 @ 22 tonnes	3	66	2.9
200	61	5 @ 22 tonnes	18	110	1.3
480	146	5 @ 22 tonnes plus 8 @ 10 tonnes	18	190	1.3
3000	915	(i) 5 @ 22 tonnes plus 8 @ 10 tonnes 72 @ 5 tonnes	18	550	0.6
		or (ii) 21 @ 24 tonnes 21 @ $1\frac{1}{2}$ tonnes	55	536	

(1) A 2 s gap at 18 mile/h speed is 55 ft
     and at 6 mile/h speed is 18 ft

It may be seen that up to 23 m a jam situation was envisaged with heavy and compact lorries. Between 23 and 61 m a column of heavy and compact lorries with 18 ft gap between them was assumed. Beyond 61 m, a further dilution by lighter vehicles was also assumed.

During the drafting of BS 5400 Part 2[2] in the 1970s, two main changes were made, namely:

(1) A horizontal cut-off at 30 kN/m was introduced for loaded length less than 30 m, combined with a requirement that all bridges carrying public highways should be designed for at least 25 units of HB loading.

(2) A horizontal cut-off was also introduced at 9 kN/m for loaded lengths greater than 380 m.

In the United States, the AASHTO Specification[5] stipulates five classes of truck loading and equivalent lane loading, both of which have to be considered in design. The truck loadings HS20 and HS15 consist of a tractor and semi-trailer combination with a variable spacing V between the rear two axles, and H20, H15 and H10 consist of a two-axle truck, as shown in Figs 3.1(a) and (b), respectively. For V the most critical value in the range 4.27 m to 9.14 m is to be taken. The equivalent land loadings consist of a uniformly distributed load together with a concentrated load, as given in Table 3.2.

For continuous spans, a second concentrated load of the same magnitude is to be considered for the calculation of hogging moments. Bridges supporting

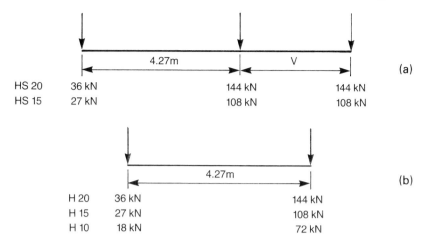

Fig. 3.1   Standard trucks for American loading.

Table 3.2   Land loading.

	Uniform load (kN/m)	Concentrated load (kN)	
		for bending moment	for shear
HS20 and H20	9.4	80	116
HS15 and H15	7.1	60	87
H10	4.7	40	58

interstate highways are designed for HS20 loading and also an alternate military loading consisting of two axles 1.22 m apart, each weighing 108 kN. For trunk highways or other highways carrying heavy truck traffic, the minimum live load is HS15. For orthotropic steel decks, instead of the 144 kN single axle in HS20 and H20 loading, one single axle of 108 kN or two axles of 72 kN each spaced 1.22 m apart may be used. For HS15, H15 and H10 loading, a provision for infrequent heavy load is made by increasing the truck loading by 100%, without concurrent loading on any other lanes, with stresses resulting from dead, live and impact loading allowed to be up to 150% of the normal allowable stresses. The lane loading of the standard truck is assumed to be 3.05 m wide and is placed in 3.66 m wide traffic lanes. Where loading in several lanes produces the maximum stress in any member, the improbability of all the lanes being loaded simultaneously is taken into account by applying the following reduction factor on the stresses:

* one or two lanes loaded   – 100%
* three lanes loaded   – 90%
* four or more lanes loaded – 75%

To allow for dynamic, vibratory and impact effects, live load stresses in superstructures due to H or HS loading are increased by an impact factor I given by

$$I = \frac{15.24}{L + 38}$$

where L is the loaded length in metres of the portion of the span.

In West Germany, DIN 1072[6] specifies three levels of design loading for three classes of bridges: class 60 for bridges on motorways, federal and state (Länder) roads, class 12 for bridges on roads for light traffic (i.e. in rural areas) and class 30 for all other bridges, i.e. on country, community and city roads and rural roads carrying heavy traffic. The carriageway is divided into traffic lanes of 3 m width; the most critical lane for the design of a structural element is called the principal lane. For the principal lane, a heavy vehicle Q and a uniformly distributed loading q_1 in front and behind this vehicle are prescribed; all the other lanes are loaded with a uniformly distributed load q_2. The values of Q, q_1 and q_2 are given in Table 3.3.

The loading on the principal lane is multiplied by an impact factor k varying between 1.0 to 1.4 and given by $k = 1.4 - 0.008L$, but not less than 1, where L is the span length of the member in metres.

In France[7], bridges are classified according to the carriageway widths:

- Class I – bridge with carriageway width equal to or greater than 7 m
- Class II – bridge with carriageway width between 5.5 and 7 m
- Class III – bridge with carriageway width equal to or less than 5.5 m.

The carriageway width is divided into an integer number of traffic lanes of width not less than 3 m, except that carriageways with widths between 5 and 6 m are considered to have two lanes. Two different and independent types of loading are considered – a uniformly distributed load A and a vehicle or axle load B. A is given by

$$A = 2.3 + \frac{360}{L + 12} \text{ kN/m}^2$$

Table 3.3 Loading of principal and other lanes.

	Heavy vehicle			Distributed load	
Class	Total load Q (kN)	Axle load (kN)	Distance between axles (m)	q_1 (kN/m^2)	q_2 (kN/m^2)
60	600	200	1.50–1.50	5	3
30	300	100	1.50–1.50	5	3
12	120	40/80	3.0	4	3

Table 3.4 Coefficients a_1.

Number of loaded lanes		1	2	3	4	5
	I	1	1	0.9	0.75	0.7
Bridge class	II	1	0.9	—	—	—
	III	0.9	0.8	—	—	—

where L = loaded length in metres. A is multiplied by coefficient a_1, which depends on the bridge class and number of lanes to be loaded, as shown in Table 3.4.

There is another multiplying factor $a_2 = V_0/V$, where V is the width of the lane being considered and $V_0 = 3.5$ m for class I, 3.0 m for class II and 2.75 m for class III. The load $(a_1\, a_2\, A)$ is placed uniformly over the total widths of the traffic lanes considered.

The vehicle or axle load B for each bridge member consists of three independent loading systems:

(1) B_c consists of two vehicles of 300 kN on each lane, with axle spacings as shown in Fig. 3.2. The value of the vehicle load is multiplied by a coefficient that depends upon the number of loaded lanes and bridge class shown in Table 3.5.

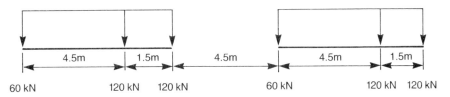

Fig. 3.2 Axle loading for French loading.

The possibility of the vehicles in two adjacent lanes being very close to each other is allowed for by considering the disposition shown in Fig. 3.3.

(2) B_r consists of an isolated wheel load of 100 kN with contact area 0.3 m along the direction of travel and 0.6 m across.

Table 3.5 Coefficients multiplying B.

Number of loaded lanes		1	2	3	4	5
	I	1.20	1.20	0.95	0.80	0.7
Bridge class	II	1.0	1.0	—	—	—
	III	1.0	0.8	—	—	—

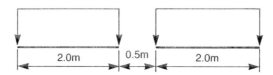

Fig. 3.3 Lateral vehicle disposition for French loading.

(3) B_t consists of a pair of two axles, each 160 kN, on each lane. The spacing between the two axles is 1.35 m and the transverse distance between the wheels is 2.0 m; the possibility of the axles in adjacent lanes being very close to each other is allowed for by taking the minimum space between the wheels of the two axles as 1.0 m. This loading is multiplied by 0.9 for bridge class II and is not considered for bridge class III.

Certain classified routes are designated for the passage of heavy military vehicles weighing up to 1100 kN or exceptional heavy transport represented by two carriers each weighing up to 2000 kN.

The impact factor is already included in the loading system A; for the loading system B, the impact factor K is given by

$$K = 1 + \frac{0.4}{1 + 0.2L} + \frac{0.6}{1 + 4P/S}$$

where P = permanent load, S = live load B, and L = length of bridge member in metres.

A study group set up by the Organisation of Economic Co-operation and Development (OECD) has produced[1] a comparative analysis of the bridge loading standards in the member countries, i.e. the bridge design loading in Belgium, Finland, France, W. Germany, Holland, Italy, Japan, Norway and Sweden, Spain, the UK and the USA are compared. In some of these countries all bridges on public roads are designed for the same loading, whereas in the other countries, the design loading depends on the type of route. Most codes prescribe the simultaneous action of a single concentrated or truck loading, and a uniformly distributed load. Some codes consider the effect of a heavier axle or wheel load. The impact effect is already included in the design loads in some codes, but in the other codes the design loading has to be increased by a factor which generally decreases with the length of the member. The intensity of loading decreases with the increase in the loaded length in most of the codes. Most codes also allow a reduction when several traffic lanes have to be loaded.

This study also included a valuable numerical exercise of calculating the total bending moments caused by the live loads of the various codes on a simply supported bridge. Separate calculations were made for the bridge carrying two, three and four traffic lanes and spanning 10–100 m. The total load on the whole bridge was considered for this comparison, as if the bridge was supported by one single beam. The impact factor and the reduction due to multiple land loading were taken into account; any difference between the various codes on the allowable stress levels was also allowed for by multiplying the bending moment by a ratio:

$$\frac{\text{yield stress of steel specified in the national material specification}}{\text{allowable stress of steel in the bridge design code}}$$

The bending moment M thus obtained was converted into an equivalent uniformly distributed load q_{eq} in kN/m, given by $\frac{8M}{L^2}$. These q_{eq} values indicate the structural strength of bridges built according to the loading specifications of the various countries. Fig. 3.4(a) and (b) show these q_{eq} values for bridges with two and four lanes in the carriageway, respectively. Very wide differences between different countries are evident, the AASHTO loading being by far the lightest for spans over 25 m.

3.4 Recent developments in bridge loading

In recent years it has been found in several countries that the standard loading does not satisfactorily reflect the effect of a long queue of vehicles in a traffic jam situation, particularly for long loaded lengths of, say, over 40 m. In the United States, proposals[8] have been made for a new loading standard which will consist of a uniformly distributed load U and a concentrated load P for each lane, both of which depend on the length of the bridge to be loaded for the worst effect. U depends also on the percentage of heavy goods vehicles in the traffic. Table 3.6 gives typical values. No allowance for impact needs to be added, as the loading represents a static jam situation. In multiple lanes, a second lane shall have 70% of the basic lane load and all other lanes shall each have 40%.

Table 3.6 Typical values of U and P for various bridge lengths.

Loaded length (m)	P (kN)	U (kN/m) for percentage of HGV in traffic		
		7.5%	30%	100%
15.25	0	38	38	38
122	320	10.4	13.9	17.1
488	534	7.1	10.8	12.3
1950	747	5.8	9.9	10.5

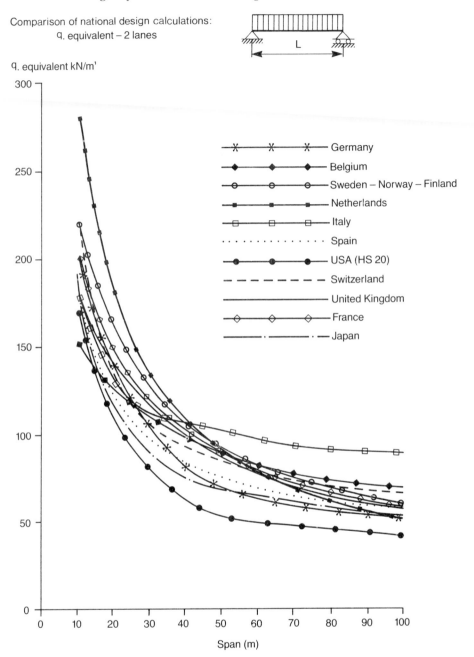

Comparison of national design calculations:
q. equivalent – 2 lanes

q. equivalent kN/m'

✕ ─── ✕ ─── ✕ Germany
◆ ─── ◆ ─── ◆ Belgium
⊖ ─── ⊖ ─── ⊖ Sweden – Norway – Finland
■ ─── ■ ─── ■ Netherlands
▱ ─── ▱ ─── ▱ Italy
· · · · · · · · · · · · Spain
● ─── ● ─── ● USA (HS 20)
─ ─ ─ ─ ─ ─ ─ Switzerland
─────── United Kingdom
◇ ─── ◇ ─── ◇ France
───·───·── Japan

Span (m)

Fig. 3.4(a)   Comparison of National Bridge Loadings for two-lane bridges. (*Courtesy OECD.*)

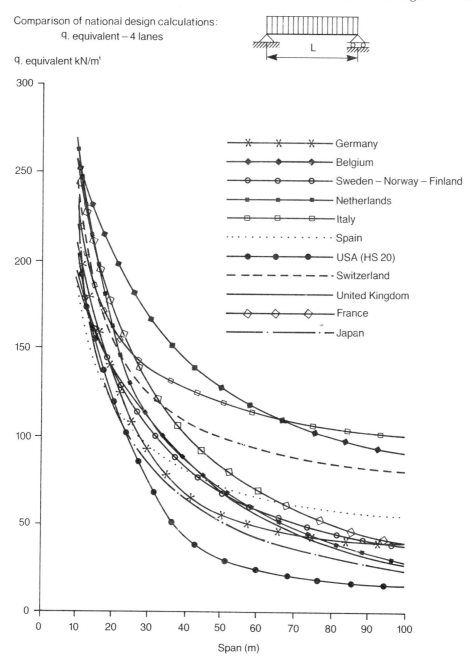

Comparison of national design calculations:
q. equivalent – 4 lanes

q. equivalent kN/m'

Germany
Belgium
Sweden – Norway – Finland
Netherlands
Italy
Spain
USA (HS 20)
Switzerland
United Kingdom
France
Japan

Span (m)

Fig. 3.4(b) Comparison of National Bridge Loadings for four-lane bridges. (*Courtesy OECD.*)

In W. Germany, a new draft bridge loading standard has been proposed in which, in addition to the heavy vehicle of 600 kN on the principal lane a heavy vehicle of 300 kN is also taken on a second lane, and the class 12 loading for lightly used roads has been abandoned. In Britain, observations on traffic queues and jams in the early 1980s raised the question whether the assumptions about gaps between vehicles and dilution by light traffic are valid for modern traffic conditions. There have also been some remarkable changes in the freight transport pattern in recent years. For example, between 1962 and 1977, goods moved by road, measured by tonne-km, virtually doubled. Statistics of vehicle population by gross weight indicate that the number of vehicles with gross weight greater than 28 tonnes increased from an insignificant number in 1962 to 90 000 in 1977. Vehicle population with gross weight less than 11 tonnes fell substantially in that period, whereas those between 11 and 28 tonnes also increased, but the rate of increase of over 28 tonnes completely outstripped that of the others.

This is the context in which a thorough review of design for live loading of bridges was undertaken in early 1980s. These reviews took into account the growth of traffic and change in traffic mix predicted for the 1990s, and produced new proposals for loading for the entire range of loaded lengths. For the shorter spans, the review was primarily on a deterministic basis, although an element of probability for illegal overloading and lateral bunching was taken into account. The extreme loading obtained from this part of the exercise was considered to be just possible in a rare event, i.e. to be used as design load in the ultimate limit state without multiplying by any further partial safety factor. The results were thus divided by 1.5 to get nominal loading. For the longer spans, a fully statistical basis was used to derive a characteristic loading, i.e. 95% probability of not being exceeded in 120 years. The nominal loading for design, i.e. a 120 yr return period load, was taken as the characteristic loading divided by 1.2. A level 3 overall safety analysis was then performed to obtain an appropriate partial safety factor for loading consistent with the assurance that a structure will not have more than 1 in a million chance of collapse in service in its lifetime. The partial safety factor on the nominal load was found to be approximately 1.5, thus confirming the value already in BS 5400. On top of the nominal design loading derived in this fashion, a 10% allowance for future contingencies was provided throughout the whole span range. On a multi-lane bridge deck the loading over the third lane onwards was increased from 33% to 60% of the above design load in the first two lanes.

In 1986 the European Commission decided to adopt a single regulation for maximum vehicle weights and dimensions throughout all the countries belonging to the European Economic Community. For Great Britain this meant an increase in the maximum axle weight from 10.5 tonnes to

11.5 tonnes, and in the maximum vehicle weight from 38 tonnes to 40 tonnes. The increase in the maximum axle weight required a slight increase in the previously derived design HA loading for the short spans. The increase in the maximum vehicle weight was, however, offset by an increase in the minimum axle spacing, thus requiring no further increase in the design loading.

The proposed new HA loading[9] is shown in Fig. 3.5 on which are also shown the HA loading in BS 153[4] and BS 5400[2]. The increase for all loaded lengths is obvious, but in the middle range of 25−60 m it is not very substantial. For the shorter loaded lengths, the apparently substantial increase is not in fact critical, since 30 units of HB loading produces worse loading effects, even after allowing for the appropriate partial factors. For loaded lengths above, say 60 m, dead load starts to become more dominant than live load, and hence the total increase in loading will not be as much as indicated in this graph.

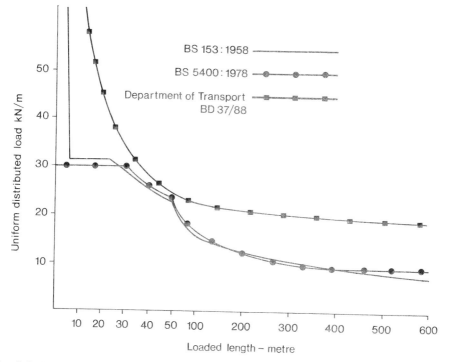

Fig. 3.5   British bridge loading.

3.5 Longitudinal forces on bridges

Longitudinal forces are set up between vehicles and the bridge deck when the former accelerate or brake. The magnitude of the force is given by

$$\frac{W}{g}\frac{\delta V}{\delta t}$$

where W is the weight of the vehicle, g is the acceleration due to gravity ($=9.81$ m/s^2) and δV is the change in speed in time δt.

Usually the change in speed is faster during braking than while accelerating; for example, lorries take over 15 s to reach 60 mile/h ($=27$ m/s) but may come to a stop in 5 s from this speed. Assuming a constant deceleration in the time to stop, the longitudinal force from one vehicle may thus reach

$$\frac{W}{9.81}\frac{27}{5} = 0.55W$$

For a 300 kN lorry, the longitudinal force may thus be as high as 165 kN. The possibility of more than one vehicle braking at the same time on a multi-lane long bridge should also be considered.

In the United States[5] the effect of braking or acceleration is taken as a longitudinal force equal to 5% of the live load in the lane specified for the maximum bending moment. This loading is taken to act at a level 1.83 m above the road surface. All lanes that may carry traffic in the same direction are to be considered. In Britain[2], the longitudinal force is taken in one notional lane only, and is equal to 8 kN/m of loaded length plus 200 kN, but not more than 700 kN, for HA loading. For HB loading, 25% of the HB load is taken as the longitudinal force. This loading is assumed to be applied on the road surface. Table 3.7 indicates the wide difference in the horizontal loads obtained from the two codes. The new American proposal[8] stipulates 80% of the design truck to be taken as the horizontal load on one lane and 5% of the lane load (including concentrated load for bending moment) in all other lanes in the same direction. This will increase the total horizontal load very substantially, bringing the value for HS20 loading nearer to the British value. The German code[10] specifies a horizontal load of 5% of the distributed loading q_1 on the

Table 3.7 Horizontal loads in British and American Standards.

	Total horizontal load (kN)				
	British Standard BS 5400		AASHTO 1977, with no. of lanes in one direction		
Loaded length (m)	HA loading	45 units HB loading	2	3	4
20	360	450	32.4	43.7	48.6
50	600	450	55.0	74.3	82.5
100	700	450	102.0	137.7	153

carriageway up to a loaded length of 200 m but not less than 0.3 times the design heavy vehicle Q (see Section 3.3); but in the new proposals, this is being increased to 10% of the distributed loading q_1 on the carriageway up to 12 m wide and 200 m long, but not less than $\frac{1}{3}$ of the design heavy vehicles Q in the main and second lanes.

3.6 Wind loading

Wind load on a bridge may act:

(1) horizontally, transverse to the direction of the span
(2) horizontally, along the direction of the span
(3) vertically, upwards causing uplift.

Wind load is not generally significant for short-span bridges; for medium spans, the design of the sub-structure may be affected by wind loading; the superstructure design is affected by wind only for long spans.

For bridges with high natural frequency of vibration, only the static loading effect of wind needs to be considered. The dynamic effect of wind, and the oscillation caused by it, is, however, very important for bridges with low natural frequency.

In the AASHTO Code[5] the transverse wind load is specified as 3.6 kN/m^2 for trusses and arches and 2.4 kN/m^2 for girders and beams, to be applied to the total exposed area in elevation. The following minimum values are also specified, corresponding to wind velocity of 161 km/h:

(1) 4.4 kN/m in the plane of the loaded chord of a truss
(2) 2.2 kN/m in the plane of the unloaded chord of a truss
(3) 4.4 kN/m on girder spans.

All these values may be reduced if a lower velocity is appropriate to the particular location. These values are appropriate to the bridge carrying no traffic load. When traffic is present, only 30% of the above load on the structure is taken, plus a loading of 1.5 kN/m acting at 1.83 m above the deck. For the effect on sub-structures, forces in both transverse and longitudinal directions are to be taken, their magnitudes depending on the angle of skew. For zero skew angle, the transverse forces are as mentioned above and the longitudinal forces are zero; the former decreases and the latter increases with increase in the skew angle. The wind loading on the sub-structure itself is taken as 1.92 kN/m^2, corresponding to 161 km/h wind speed, and the wind force is resolved into transverse and longitudinal components corresponding to the skew angle of the wind direction. When traffic is present, only 30% of the wind loading on superstructure and sub-structure is considered along with the full wind loading on the traffic. An uplift force of 1 kN/m^2 on the loaded bridge on the total plan area is considered for overturning effects.

In the British Standard BS 5400[2], isotachs of 120 yr return mean hourly wind speeds in the British Isles at 10 m height in open country are provided. This hourly wind speed at the particular site is then adjusted to obtain the maximum gust speed V_c, allowing for:

(1) the height of the bridge above ground level
(2) gusting effect over a 3 s period
(3) non-coherence of gusting over the length of the bridge
(4) possible local funnelling or acceleration of wind at bridge site.

When a live load is present on the bridge, V_c is limited to 35 m/s. A factor of 1.1 is taken for sites susceptible to (4) above. The other three parameters are dealt with by a factor S_2 given by[11]

$$S_2 = \left(\frac{Z}{10}\right)^{0.17} \left[1 + I^2 + 2\sqrt{2}\, pI \left\{\frac{l}{L} - \frac{l^2}{L^2}(1 - e^{-\frac{L}{l}})\right\}^{\frac{1}{2}}\right]^{\frac{1}{2}}$$

where Z = height in metres above ground level
      L = horizontal length subjected to wind
      I = intensity of turbulence

$$= 0.18\,[1 - 0.00\,109\,(Z-10)]\left(\frac{10}{Z}\right)^{0.17}$$

   l = crosswind width of gust
     = 50 m for $Z > 50$ m
     = $0.375Z + 31.25$, for $Z \le 50$ m
   P = peak force factor

$$= \frac{1}{2I}\left[2.56\left(\frac{10}{Z}\right)^{0.17} - 1 - I^2\right]$$

Transverse wind loading is then calculated as

$$\tfrac{1}{2}\rho\, V_c^2\, A_e\, C_D$$

where ρ is the air density 1.226 kg/m^3, V_c is the gust speed in m/s, A_e is the area in elevation in m^2, and C_D is the drag coefficient.

For plate or box girder types of bridges, the area in elevation includes a 2.5 m high vertical surface above the bridge deck for wind loading on vehicles; any solid parapet is also included, but wind loads on open parapets are calculated separately. For truss bridges, wind loads are cal-culated separately for:

(1) windward and leeward girders
(2) windward and leeward parapets
(3) area in elevation of the deck structure
(4) live load height 2.5 m above deck.

Any screening offered by an adjacent component can be taken advantage of.

For plate and box girder types of bridges, the drag coefficient C_D is dependent upon the ratio of the width b of the bridge to its maximum depth d as seen in the bridge cross-section. The value of C_D varies from 1.4 for a b/d ratio of 4, to 1.0 for a b/d ratio equal to or greater than 12. For a very narrow and deep bridge cross-section with a b/d ratio of 0.6, C_D reaches a peak value of 2.75. For truss girder bridges with flat-sided members, C_D for the windward truss is dependent upon the solidity ratio, i.e. the net area to the overall area of the truss in elevation; C_D varies from 1.9 to 1.6 for solidity ratio 0.1 to 0.5. For trusses with round members, C_D for the windward girder is 1.2 or 0.8 depending on whether gust speed × diameter < or ⩾6 m²/s. For the leeward truss, this drag coefficient is multiplied by a shielding factor which depends on both the solidity ratio defined above and the spacing ratio, i.e. the distance between the trusses divided by their depth. The shielding factor varies from 1.0 for the solidity ratio of 0.1 and any spacing ratio, to 0.45 for solidity ratio 0.5 and spacing ratio 1, and to 0.7 for solidity ratio 0.5 and spacing ratio 6. The drag coefficient for unshielded parts of the live load is taken as 1.45. Tables are given for C_D for different shapes of parapets and bridge piers.

The longitudinal wind load on superstructures is taken as one-quarter of the transverse wind load on plate or box girder bridges, one-half of the transverse wind load on trusses, and also one-half of the transverse wind load on vehicles. Upward or downward wind load is taken as

$$\tfrac{1}{2}\rho\, V_c^2\, A_p\, C_L$$

where ρ is the air density = 1.226 kg/m³, V_c is the gust speed in m/s, A_p is the plan area in m², and C_L is the lift coefficient.

For bridges with deck superelevation up to 1°, C_L is 0.4 for b/d ratio up to 7 and 0.15 for b/d ratio greater than 16, with a linear variation in between; for superelevation between 1° and 5°, C_L is 0.75.

A combination of full transverse and vertical loading due to wind is considered, but full longitudinal loading is considered in combination with half the transverse and vertical loading.

The new American bridge loading proposals[8] are similar to, but probably not as detailed as, the British loading. The transverse wind loading on the full area in elevation is stipulated as

$$\frac{Z^{0.2} \, V_{30}^2 \, C_d}{600} \text{ lb/ft}^2$$

where Z is the height in feet of the bridge deck surface above ground or water level but not less than 30 ft, V_{30} is the 100 yr return fastest mile wind speed in miles per hour at the 30 ft height and can be obtained from the map of isotachs or preferably from local wind data, and C_d is the drag coefficient and is specified as 1.5 for plate or box girder bridges and 2.3 for truss bridges unless a lower value is justified by wind tunnel tests. When a live load is present on the bridge, a vertical surface of 10 ft height, less the area shielded by solid parapets, is included in the total area in elevation, C_d is specified as 1.2 on this part, and V_{30} is limited to 55 mile/h. The vertical wind load on the plan area of the bridge is stipulated as

$$\frac{Z^{0.2} \, V_{30}^2 \, C_L}{600} \text{ lb/ft}^2$$

where C_L is the lift coefficient specified as 1.0.

The above two proposed formulae for wind load are based on the following assumptions:

(1)   The maximum gust speed is approximately 1.6 times the mean hourly wind speed, but because of the incoherence of gusts along the whole length of the bridge the gust speed is reduced to 1.41 times the mean hourly speed.
(2)   At heights above 30 ft the gust speed is assumed to increase according to the $\frac{1}{10}$th power of height.
(3)   The mean hourly wind speed is 0.8 times V_{30}, the fastest mile wind speed at 30 ft, in which form the wind speed is recorded in the United States.
(4)   The maximum gust speed, unlike the mean hourly speed, is relatively insensitive to terrain condition.

At 30 ft height and $V_{30} = 100$ mile/h, the new formulae give very similar wind loads to the current AASHTO values, but for other heights and locations the new values will obviously be different. At $Z = 30$ ft and $V_{30} = 80$ mile/h, the horizontal wind pressure given by the new loading is 21 lb/ft^2, which can be compared with the values of 23 and 19.5 lb/ft^2 given by the British Standard for mean hourly wind speed of 64 mile/h, 10 m height and bridge lengths 60 m and 200 m, respectively. The proposed new American drag coefficients of 1.5 and 2.3 for solid and truss girders may be compared with the British values of 1.4 to 1.0 for solid girders with b/d ratios ranging from 4 to 12, and 1.6 on the windward truss and 0.7 on the leeward truss with flat-sided members, with a solidity ratio of 0.5 and a shielding factor of 0.5. The lift coefficient of 1.0 for vertical wind load in the American proposals can be compared with the British value of 0.75 for superelevation between 1 and 5 degrees. With traffic present on the bridge the American proposal specifies $V_{30} = 55$ mile/h, which is equivalent to a gust speed of $1.41 \times 0.8 \times 55 = 62$ mile/h or 27.6 m/s, but with the wind load dependent upon the height of the bridge deck above ground or water level; this may be compared with the gust speed of 35 m/s for all heights stipulated in the British code.

In the German code[6], wind load is specified as 2.5 kN/m² without traffic, and 1.25 kN/m² with traffic, to be applied to the area in projected elevation of the bridge. The traffic profile is taken as a 2 m high vertical surface above the bridge deck. The proposed new German loading[10] retains the above wind loading for bridges with superstructure 50 to 100 m above ground level, but makes reductions for:

(1)   superstructures at lower height
(2)   superstructures with noise barriers, in the load case without traffic.

It also increases the height of the traffic profile to 3.5 m.

3.7   Thermal forces

If the free expansion or contraction of a structure due to changes in temperature is restrained by its form of construction (e.g. portal frame, arch) or by bearings or piers, then stresses are set up inside the structure. Secondly, differences in temperature through the depth of the superstructure set up stresses if the structure is not free to deform. A differential temperature pattern in the depth of the structure represented by a single continuous straight line from the top to the bottom surface does not cause stresses in a statically determinate structure, e.g. simply supported or balanced cantilever spans, but will cause stresses in a continuous structure due to the vertical restraints provided by the piers. Normally differential temperature is not represented by a single continuous line from the top to the bottom surface, and hence causes stresses even in simple spans.

In the British Standard BS 5400[2], maps of isotherms provide the extremes of shade air temperatures at sea level in different parts of the British Isles. For heights above sea level these temperatures are reduced by 0.5°C and 1.0°C for minimum and maximum temperatures, respectively, for every 100 m height. Local peculiarities like frost pockets, sheltered areas, urban or coastal sites should also be taken into account. The minimum temperature in the bridge structure is usually lower than the

minimum shade air temperature by 2 to 4°C for bridges with steel orthotropic decks and higher by 1 to 8°C for bridges with concrete decks; the
maximum bridge temperature is higher than the maximum shade air
temperature by between 9 and 20°C for bridges with steel decks and by up
to 11°C for bridges with concrete decks. The difference between the
bridge and the shade air temperatures depend upon the latter and also on
the type and depth of surfacing provided on the bridge deck; data for
these differences are tabulated in the British code[2]. Within this range of
the bridge temperatures, the variation with respect to the particular datum
temperature at which restraint was imposed on the bridge during its
construction determines the magnitudes of thermal stresses.

In the AASHTO code[5], a range of bridge temperatures of −18° to
+49°C is specified for a moderate climate and −34° to +49°C for a cold
climate.

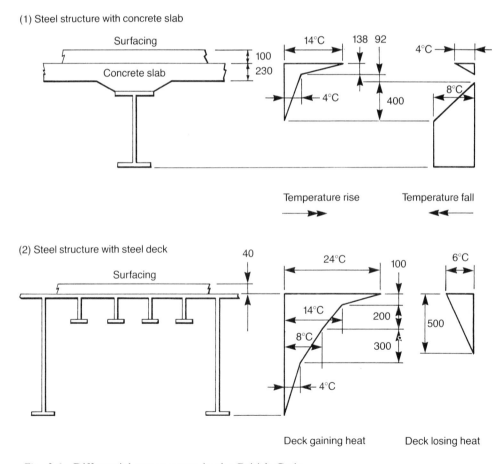

Fig. 3.6   Differential temperature in the British Code.

The differential temperature pattern given in the British code[2] is based on extensive measurements on bridges in the British Isles and deals with various types of bridge decks and deck surfacings. For the common case of a steel plate or box or truss girder construction with (1) a 230 mm thick concrete slab and 100 mm of deck surfacing and (2) a steel orthotropic deck with 40 mm of surfacing, the temperature differential with the road surface in the hot and cold conditions are as shown in Fig. 3.6.

The AASHTO code[5] does not specify any temperature differential, but the new proposals[8] stipulate the pattern shown in Fig. 3.7.

In the German code[6], the temperature at the time of construction is assumed to be +10°C and a variation of ±35°C from the construction

(1) Steel structure with concrete slab

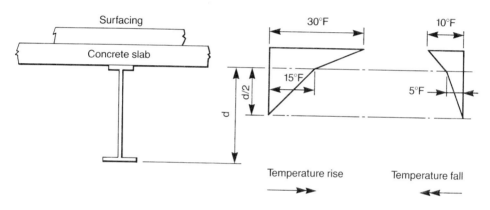

(2) Steel structure with steel deck

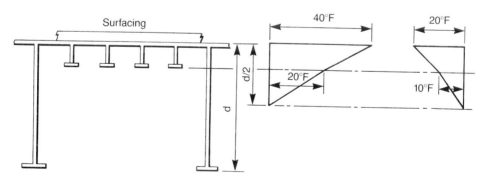

Fig. 3.7 Differential temperature proposed in the USA.

temperature is to be considered; within this range, a differential temperature of 15°C, linearly varying between different parts of the bridges structure, is also to be considered, for example between top and bottom flanges, between cables and stiffening girders, between webs of box girders.

In composite structures, i.e. steel structures with concrete slabs, a temperature increase or decrease in the top surface of the slab of 20°C and at the bottom edge of the steel girder of 35°C, from the construction temperature of +10°C, is specified. In the new German proposals[10] clarifications have been made that differential temperature need only be considered in the vertical plane and the magnitudes have been proposed to be reduced to 10°C with the deck hot, 5°C with the deck cold for a steel deck bridge and 7°C with the deck cold for a composite bridge with a concrete deck. With traffic load on the bridge, either differential temperature or the traffic load may be reduced to 70%. A temperature difference of ±15°C between different members of a bridge that are generally unconnected to each other should also be considered, for example between the beam and arch, cables and deck structure, upper and lower chords of trusses.

3.8   Other loads on bridges

There are several other sources for loads or stresses in bridge structures, namely:

(1)   centrifugal forces on a horizontally curved bridge
(2)   accidental load due to skidding or collision with parapet
(3)   creep and shrinkage of concrete
(4)   snow load on bridge deck, cables, etc.
(5)   friction at, or shearing resistance of, bearings
(6)   earth pressure on retaining structures
(7)   stream flow pressure, floating ice, buoyancy
(8)   earthquake or ground movement due to other causes
(9)   settlement of supports
(10) impact from shipping.

National standards provide relevant data for these loadings. Loading due to (1), (2), (3), (5) and (6) should not vary a great deal from country to country, but the different national standards stipulate very different values. The other loadings are dependent upon the geographical conditions of the bridge site and thus vary widely from country to country.

3.9 Load combinations

The concept of (1) a nominal or characteristic value of a load and (2) a partial safety factor by which this nominal or characteristic value is multiplied to obtain the design value of the load is discussed in Chapter 4. When several loads are to be combined, the partial safety factors should be reduced from their values for individual application of the loads in order to attain the same probability of occurrence of the combination as that of the individual loads.

In the 'permissible stress' method of design (see Chapter 4) a specified percentage of overstress was allowed for the total stress due to several loads acting in combination.

The load combinations specified in the British code[2] are five in number. Dead load of the bridge structure and the superimposed dead load from deck surfacing, parapet, etc. and earth pressure are permanent loads and are included with specified load factors in all combinations. A high load factor is taken for superimposed dead load, in order to allow for further resurfacing with incomplete removal of previous layers; but a reduced value may be taken when precautions are taken against this occurring. Load combination 1 is for permanent loads and the main traffic loading. Load combination 2 is for wind loading: 2(a) for wind without traffic and 2(b) for wind with traffic. Load combination 3 is for maximum thermal effects. Combination 4 is for secondary and accidental traffic loading. Combination 5 is for friction forces at the bearing; it is considered that vibration effects due to live loads overcomes frictional forces and thus the live load need not be taken together with frictional forces. The load combinations and the partial load factors for the ultimate limit state are given in Table 3.8.

The permanent loads and their factors are not mentioned in the Table; these are:

1.05   for dead load of steelwork
1.15   for concrete
1.75   for superimposed dead load, but may be reduced to 1.20
1.50   for earth pressure
1.20   for creep and shrinkage of concrete

The different design parts of the British Standard BS 5400 allow the effects of differential temperature, differential settlement and creep and shrinkage effects of concrete to be ignored in the ultimate limit state in those types of structural design that are capable of redistributing the self-equilibrating internal stresses set up by these effects. Snow load, loading

Table 3.8 Load combinations and partial load factors for ultimate limit state.

Combination		LL[1]	LF	CF	CL	W	T	F
1		1.5	–	–	–	–	–	–
2	(a)	–	–	–	–	1.4	–	–
	(b)	1.25	–	–	–	1.1	–	–
3		1.25	–	–	–	–	1.3[2]	–
	(a)	1.5[3]	–	1.5	–	–	–	–
4	(b)	1.25[3]	1.25	–	–	–	–	–
	(c)	1.25[3]	–	–	1.25	–	–	–
5		–	–	–	–	–	–	1.3

Load combination in British Standard BS 5400

LL = live load
LF = longitudinal force
CF = centrifugal force
CL = collision force
W = wind loading
T = temperature effects
F = bearing friction

(1) factors for HA loading
(2) load factor for differential temperature effects is 1.0
(3) these factors are used on reduced live loading.

due to stream current and floating ice, earthquake and shipping collision forces are not included in the specified load combinations, as few bridges in the British Isles are subjected to these loadings; where such loading is likely special load combinations and partial load factors are adopted.

In the AASHTO specification[5], nine load combinations are specified. Each load is multiplied by one γ-factor particular for the load combination under consideration and one β-factor which varies for different loads in the same combination. Thus the total load effect in combination N is given by

$$\gamma_N \Sigma(\beta_{FN} F)$$

where F represents the load due to dead load, live load, wind load, etc. and β_{FN} is the factor on the particular load for the particular combination N. Dead load, earth pressure, buoyancy and stream flow pressure are considered permanent forces and are included in all the load combinations with the following β-factors:

- Dead load   = 1.0 except that a smaller factor of 0.75 is taken for minimum load and maximum moment in columns
- Earth pressure = 1.3 for maximum lateral pressure
              = 0.5 for minimum lateral pressure
              = 1.0 for vertical earth pressure

- Buoyancy $= 1.0$
- Stream pressure $= 1.0$

Combination 1 is the main combination with live load, combination 2 is for wind load on unloaded bridges and combination 3 is for wind load on bridges carrying traffic. Combination 4 is for temperature and shrinkage effects with live load and combination 5 is for temperature and shrinkage effects with high wind and no live load. Combination 6 is an omnibus combination with all the above loads, but with a reduced γ-factor to represent the reduced likelihood of all the forces acting with their peak values. Combinations 4, 5 and 6 are dominated by the forces due to shrinkage and thermal effects and are thus critical for those structures that are restrained against longitudinal expansion or contraction, i.e. arches and portal frames. Combination 7 is for earthquake forces, to be taken in conjunction with only the permanent loads. Combinations 8 and 9 are for ice pressure on substructures, to be combined with live loads only in the former combination, and with wind load only in the latter.

In comparing the γ and β factors in the AASHTO specification with the partial load factors in BS 5400, it has to be remembered that in the latter code two other partial factors are to be considered. These are:

(1) γ_{f3} — this takes account of inaccuracies in the assessment of load effects, or in the calculation model or in the overall dimensions, and is taken as 1.1 in the ultimate limit state.
(2) γ_m — this is the partial factor for material strength and is generally 1.05 for structural steel.

3.10 References

1. Evaluation of Load Carrying Capacity of Bridges. Organisation for Economic Co-operation and Development, Paris (1979).
2. BS 5400: Part 2: 1978. Steel, Concrete and Composite Bridges: Specification for Loads: British Standards Institution, London.
3. Departmental Standard BD 23/84: Loads for Highway and Foot/Cycle Track Bridges. Department of Transport, London.
4. British Standard 153: 1958. Specification for Steel Girder Bridges. British Standards Institution, London.
5. Standard Specifications for Highway Bridges: 12th Edition: 1977. The American Association of State Highway and Transportation Officials, Washington.
6. Deutsche Normen DIN 1072: 1967: Road and Footbridges: Design Loads. Berlin.
7. Cahier des prescriptions communes applicables aux marchés de travaux publis relevant des services de l'equipement: 1973. Ministere de l'Equipement et du Logement – Ministere des Transports, Paris.
8. Recommended design loads for bridges. Committee on Loads and Forces on Bridges of the Committee on Bridges of the Structural Division. *Journal of the Structural Division*, Proceedings of the American Society of Civil Engineers. December 1981.

9. Departmental Standard BD 37/88: Loads for Highway Bridges: Department of Transport, London.
10. Deutsche Normen. Draft DIN 1072: August 1983: Road and Footbridges: Design Loads. Berlin.
11. Hay J. S. (1974) Estimation of wind speed and air temperature for the design of bridges. Laboratory Report LR 599, Transport and Road Research Laboratory, England.

Chapter 4
Aims of Design

4.1 Limit state principle

The aim of design is that the structure should:

(1) sustain all loads and deformations liable to occur during its construction, use and also foreseeable misuse or accident
(2) perform adequately in normal use
(3) have adequate durability.

   When a structure or any of its components infringes one of its criteria for performance or use, it is said to have exceeded a limit state. For most structures the limit states can be placed in two categories:

(1) the ultimate limit states which are related to a collapse of the whole or a part of the structure
(2) the serviceability limit states which are related to disruption of the normal use of the structure.

   Ultimate limit states should have a very low probability of occurrence, since they may cause loss of life, amenity and investment. The common ultimate limit states are:

(1) loss of static equilibrium of a part or the whole of the structure considered as a rigid body (e.g. overturning, uplift, sliding)
(2) loss of load-bearing capacity of a member due to its material strength being exceeded, or due to buckling, or a combination of these two phenomena, or fatigue
(3) overall instability, leading to very large deformation or collapse, caused by, for example, aerodynamic or elastic critical behaviour or transformation into a mechanism.

The serviceability limit states depend on the function of the structures; for bridges they correspond to:

(1) excessive deformation of the structure, or any of its parts, affecting the appearance, functional use or drainage, or causing damage to non-structural components like deck joints, surfacing, etc.
(2) excessive local damage like cracking, splitting, spalling, yielding or slip, affecting appearance, use or durability of the structure
(3) excessive vibration causing discomfort to pedestrians or drivers.

4.2 Permissible stress method

In the modern generation of structural design codes, the specific requirements for the relevant limit states are stated explicitly. In the past, however, the codes did not identify the various limit states separately; they were like a cooking recipe which produced the desired end product, but the ingredients of which were not specifically chosen for particular objectives.

The process of structural design is not an exact science, nor are the data on which a design can be based accurate. There are uncertainties in the loading, in the material properties, in the engineering analysis and in the construction process. In the past, design codes allowed for these uncertainties by specifying a permissible stress for the most adverse combination of working loads. The permissible stress was obtained by applying a factor of safety on the stress observed or calculated to occur at failure. The failure stress was generally taken as the yield stress and the working loads were specified as those loads that could be expected to act on the structure several times in its design life.

In this permissible stress or working load method a structural analysis was made to evaluate the working stresses at the specified combination of working loads, which were then checked against the specified permissible stress. Thus,

$$\Sigma \text{ working stress} \leqslant \text{permissible stress}$$

$$\text{i.e.} \leqslant \frac{\text{failure stress}}{\text{safety factor}}$$

The main advantage of this method is simplicity. Because stresses, and hence deformations/deflections, were kept low under working loads, non-linearity of material and/or structural behaviour could be neglected and working stresses were calculated from linear elastic theories. Stresses from various loads could thus be added together. The disadvantages of this method are:

(1) One global factor of safety cannot deal with the different variabilities of different loads; for example, variations of dead load from the calculated working value is usually small compared with the variation of extreme wind or vehicle loads from their working values. This is particularly serious when two loads of different variabilities counteract each other; the safety factor used in this method may give a very false impression of the danger that can be caused by a modest increase in one of the loads.

(2) The analysis of the structure under working loads may not provide a realistic assessment of the behaviour of the structure at failure. If the critical components of the structure possess sufficient ductility, redistribution of load takes place after some of these components reach the limit of their linear behaviour, thus mobilising the strength of the less critical components. Formation of successive plastic hinges in a continuous beam or frame is one example. Here the safety factor in the working stress method may represent a too pessimistic ratio of the real ultimate strength of the structure to the working load. Conversely, in some structures or structural components, stresses increase faster than the loads and thus the real ultimate load is less than the working load times the safety factor.

Structures designed by the permissible stress/working load methods used to have moderate stresses in service conditions and thus the serviceability requirements like deflections, cracking, yielding, slip and vibrations were not generally critical and hence did not require checking. However, post-war developments of materials of higher strength, welding and prestressing have necessitated explicit serviceability requirements in various design codes, often in the form of empirical rules.

4.3   Limit state codes

In the modern limit state design codes, a calculation model is established for each limit state to verify that the probability of its non- exceedance is equal to or higher than a pre-defined target reliability of the structure. This model incorporates:

(1) all the possible modes in which the particular limit state may be exceeded
(2) the uncertainties of all the variable parameters involved in the model, and the uncertainty or approximation of the model itself
(3) the target reliability.

The latter is often a compromise between the initial cost of the structure and the consequences of the exceedance of the limit state, and is guided by past experiences of design and performance of similar types of structures. The variable parameters are commonly the 'actions' (i.e. forces and constrained deformations), the properties of materials, the geometrical parameters of the structure, and the inaccuracy of the model itself. In the mathematical model their variabilities can be treated by different levels of approximation. These are:

(1)   Level III — This is the most complex method; the full probability distribution of all the design variables is integrated numerically by multi-dimensional convolution integrals to compute the exact probability of failure of the structural system in all the possible modes. Because of its inherent numerical difficulties it is not suitable for design purposes except for very special structures.

(2)   Level II — In this method some idealisations are introduced into the probability analysis to reduce the numerical difficulties; thus, for each mode of failure, a failure boundary is defined by structural theories in the space of the variable parameters, from the probability distribution of the variables a checking point on the failure boundary is identified where failure is most likely to occur, and by linearising the failure boundary at the design point an approximate reliability of the structure is estimated.

(3)   Level I — This is a semi-probablistic method in which appropriate levels of reliability are achieved for each structural element by the application of a number of partial safety factors to a pre-defined set of characteristic values of the variables. The characteristic value of each variable has a pre-defined low probability of occurrence and is determined, wherever possible, from the mean value, the standard deviation and the distribution type of the variable obtained by tests or measurement. When statistical data are not available, nominal values based on past practice are used. The partial safety factors may be determined by a Level II (or III) method for the required degree of safety. Thus the Level I method can be made identical to Level II (or III) if the partial safety factors are expressed as continuous functions of the means, standard deviations and distribution types of the variables. However, most structural codes drafted in Level I format prescribe discrete values of the safety factors instead of continuous functions, to be applied to a rationalised, i.e. reduced, number of design variables.

The idea that the statistical variation in a parameter should be considered in structural design is not new. For example, the design wind speeds are determined from the distribution of the annual extreme mean hourly speeds in the British codes and of the annual extreme fastest mile speeds in North America. The acceptance criteria for the concrete mix are designed to ensure that the probability of producing concrete with a cube strength less than the specified characteristic value is less than a pre-defined target, which is 5% in the UK and 10% in the USA. Probability based limit state codes recognise that, in the presence of uncertainties, absolute reliability cannot be achieved, but the probability of exceeding a limit state can be ensured to be acceptably low.

In between the permissible stress codes and the limit state codes there have been several intermediate developments. For example, the load and resistance factor designs developed in the United States[1–3] use factored loads and factored resistances, with different factors for different loads, reflecting their different degrees of variability. Thus

$$\Sigma \text{ (Nominal Loads} \times \text{Load Factor)} \leq \frac{\text{Resistance}}{\text{Resistance Factor}}$$

This method does not deal with all the limit states, and the factors are based on past experience, intuition and perception regarding the uncertainties involved. But this is an attempt to achieve uniform safety over the range of the likely loads.

The limit state codes developed in the United Kingdom, namely CP 110[4] and BS 5400[5], prescribe all the limit states that should be considered, but the partial safety factors were again based on experience, intuition and judgement; only in the case of the design code for steel bridges in BS 5400 were the resistance factors for the main ultimate limit states determined by a Level II analysis[6].

4.4 The derivation of partial safety factors

In the concept of probability based design, a limit state is idealised as a function

$$g(x_1, x_2, \ldots, x_n) = 0$$

where the x_i are independent variable parameters like actions, material strength, dimensions, etc.; the limit state is exceeded when $g < 0$. Probability P_f that the limit state is exceeded is then calculated from

$$P_f = \int \ldots \int f_x(x_1, x_2, \ldots, x_n) \, \mathrm{d}x_1, \mathrm{d}x_2 \ldots \mathrm{d}x_n$$

in which f_x is the joint probability density function for $x_1, x_2, \ldots, x_n$, and the integration is performed over the entire space when $g < 0$.

If a limit state function can be expressed in terms of just two independent variables, i.e. a resistance variable R (e.g. the bending capacity of a beam) and an action-effect variable S (e.g. the bending moment caused by the loads), then the limit state is exceeded if $R < S$. The probability of this occurring is given by

$$P_f = \int_{-\infty}^{\infty} F(R). f(S)\, ds$$

where $F(R)$ is the cumulative probability distribution function of variable R, equal to $\int_{o}^{R} f(R)\, dR$, and $f(R)$ and $f(S)$ are, respectively, the probability density functions of the variables R and S. This is illustrated in Fig. 4.1.

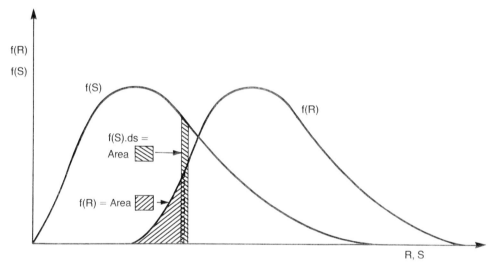

Fig. 4.1   Probability distribution of a two-variable limit state.

A safety margin Z may be defined as $Z = R - S$, $Z < 0$ representing failure. If R and S are normally distributed, Z will also be normally distributed. If the mean and the standard deviation are represented by m and σ with appropriate suffixes, then:

$$m_z = m_R - m_S$$

$$\sigma_z = \sqrt{\sigma_R^2 + \sigma_S^2}$$

The probability that $Z < 0$ is equal to the hatched area under the probability density function $f(Z)$ shown in Fig. 4.2.

If R and S are normally distributed uncorrelated random variables, the probability of failure P_f is given by

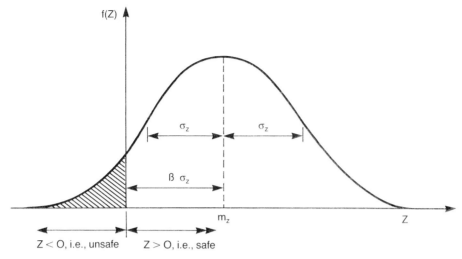

Fig. 4.2   Probability distribution of the safety margin Z.

$$P_f = \Phi \left[\frac{-m_z}{\sigma_z} \right] = \Phi \left[-\frac{m_R - m_S}{\sqrt{\sigma_R^2 + \sigma_S^2}} \right] = \Phi \left[-\beta \right]$$

where Φ is the standardised normal distribution function of cumulative densities. A reliability index β is defined as the ratio (m_z/σ_z), which is also the number of standard deviations by which m_z exceeds zero, as shown in Fig. 4.2.

If the basic variables R and S are replaced by a pair of reduced variables ω_R and ω_S given by:

$$\omega_R = \frac{R - m_R}{\sigma_R}$$

$$\omega_S = \frac{S - m_S}{\sigma_S}$$

then the failure condition $g(\omega_R, \omega_S) = 0$ is given by the equation of the straight line

$$g(\omega_R, \omega_S) = \omega_R \, \sigma_R + m_R - \omega_S \, \sigma_S - m_S = 0$$

This is shown in Fig. 4.3.

The frequency of occurrence of a particular value x of a normally distributed variable is given by the density function

$$f(x) = \frac{1}{\sigma \sqrt{2 \pi}} \, e^{-\frac{1}{2} \left(\frac{x - m}{\sigma} \right)^2}$$

where m and σ are the mean value and the standard deviation. Expressed in reduced coordinate $\omega = \dfrac{x - m}{\sigma}$, the frequency of a particular value occurring is proportional to $e^{\frac{1}{2}\omega^2}$. With two uncorrelated normally distributed variables, the frequency of occurrence of a particular set of values ω_R and ω_S will thus be proportional to

$$e^{-\frac{1}{2}\omega_R^2} \cdot e^{-\frac{1}{2}\omega_S^2} = e^{-\frac{1}{2}(\omega_R^2 + \omega_S^2)}$$

In the space of the reduced variables, the locus of the point of equal frequency will thus be a circle around the origin, larger radius representing lower frequency. These circles are shown in Fig. 4.3.

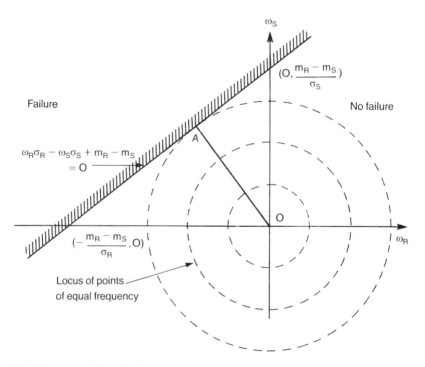

Fig. 4.3   Failure condition in the space of two reduced variables.

In Fig. 4.3 a vector OA is drawn from the origin normal to the failure boundary. OA is thus the shortest distance from the origin to the failure line. Point A represents the most likely set of values of ω_R and ω_S for the occurrence of failure and is thus called the 'design point'. It can be shown that the length of the vector OA is given by

$$\frac{m_R - m_S}{\sqrt{\sigma_R^2 + \sigma_S^2}}$$

and is thus numerically equal to the reliability index β.

The coordinates of the point A in the space of ω_R and ω_S can be shown to be:

$$\left[-\frac{(m_R - m_S)\,\sigma_R}{(\sigma_R^2 + \sigma_S^2)}, \frac{(m_R - m_S)\,\sigma_S}{(\sigma_R^2 + \sigma_S^2)} \right]$$

which can also be expressed as $(\alpha_R\beta, \alpha_S\beta)$ when

$$\alpha_R = \frac{-\dfrac{\partial g}{\partial \omega_R}}{\left[\left(\dfrac{\partial g}{\partial \omega_R}\right)^2 + \left(\dfrac{\partial g}{\partial \omega_S}\right)^2 \right]^{\frac{1}{2}}}, \quad \alpha_S = \frac{-\dfrac{\partial g}{\partial \omega_S}}{\left[\left(\dfrac{\partial g}{\partial \omega_R}\right)^2 + \left(\dfrac{\partial g}{\partial \omega_S}\right)^2 \right]^{\frac{1}{2}}}$$

The design of new structures can be performed by considering any point on the horizontal axis of R or S in Fig. 4.1 and its associated probability density values of R and S. Alternatively, the distance of the point from the mean values of R and S in multiples of the respective standard deviation may be used. For the sake of convenience, the 'design point' can be chosen for this purpose, as this point represents the most likely situation at failure.

Converting the reduced variables ω_R and ω_S to basic variables R and S, the values of the latter at the 'design point' are:

$$R_d = m_R + \omega_R\sigma_R = m_R + \alpha_R\beta\sigma_R$$
$$S_d = m_S + \omega_S\sigma_S = m_S + \alpha_S\beta\sigma_S$$

Usually the resistance and strength variables R and S of a structure are functions of a number of variables x_1, x_2, ..., x_n, and the probability density functions $f(R)$ and $f(S)$ depend upon the probability density functions of the individual variables x_1 etc. and how they are related in the functions R and S. Some of the variables may be common to both, causing some correlation; this has to be taken into account by modifying the statistical parameters by the correlation coefficient. All the basic variables may be replaced by a new set of reduced variables defined by

$$\omega_i = \frac{x_i - m_{xi}}{\sigma_{xi}}$$

where m_{xi} and σ_{xi} are the mean and the standard deviation of the basic variable x_i. The failure condition may then be written as

$$g(\omega_1, \omega_2, \ldots, \omega_n) = 0$$

This equation will represent the failure surface in the multi-dimensional space of the reduced variables ω_i. The reliability index β is the shortest distance from the origin to the failure surface. The failure surface is often curved; it is then necessary to try several points on the curved surface to obtain the shortest vector OA that is also normal to the failure surface. This is equivalent to linearising the failure surface at the design point by means of, say, Taylor's series. The coordinates of the design point are given by $\omega_i = \alpha_i \beta$,

$$\alpha_i = \left(-\frac{\partial g}{\partial \omega_i} \right) \Big/ \left[\Sigma \left(\frac{\partial g}{\partial \omega_i} \right)^2 \right]^{\frac{1}{2}}$$

calculated at the design point. α_i is dependent upon the sensitivity of the failure equation to variation in the variable ω_i in the region of the design point. α_i is proportional to σ_i in the case of planar failure surface; also, for non-planar failure surfaces, σ_i has a big influence on α_i. It may be noted that $\Sigma \alpha_i^2 = 1$.

The values of the basic variables represented by the design point are

$$x_{di} = m_{xi} + \alpha_i \beta \sigma_{xi}$$

The partial safety factor γ_i for each basic variable x_i is the ratio of its 'design' value given above to a 'nominal' value or a 'characteristic' value given in a code. A nominal value is usually based on past practice, without any statistical analysis of the probability of its occurrence. A characteristic value x_{ki} corresponds to a stipulated probability of non-compliance, usually 5%, and is determined by statistical analysis of test or measurement data; it is given by

$$x_{ki} = m_{xi} + k_i \sigma_{xi}$$

where k_i is the number of standard deviations, depending on the stipulated probability of non-compliance and the nature of the probability distribution of the variable x_i. For example, for a 95% characteristic value of a normally distributed variable (i.e. 5% probability of non-compliance) $k_i = 1.64$. Obviously, k_i is to be taken with the same sign as α_i. Thus

$$\gamma_i = \frac{x_{di}}{x_{ki}} = \frac{m_{xi} + \alpha_i \beta \sigma_{xi}}{m_{xi} + k_i \sigma_{xi}} = \frac{1 + \alpha_i \beta v_{xi}}{1 + k_i v_{xi}}$$

where v_{xi} is the coefficient of variation of the variable x_i, equal to σ_{xi}/m_{xi}.

The values of the reliability index β corresponding to various failure probabilities P_f can be obtained from the standardised normal distribution function of cumulative densities, and are given in Table 4.1.

Table 4.1 Reliability index for various failure probabilities.

β	2.32	3.09	3.72	4.27	4.75	5.20	5.61
$P_f = \Phi(-\beta)$	10^{-2}	10^{-3}	10^{-4}	10^{-5}	10^{-6}	10^{-7}	10^{-8}

The failure probability P_f required in codes for particular classes of structures should generally be derived from experiences with past practice, consequences of failure and cost considerations. Having chosen P_f, and hence β, the determination of the appropriate partial safety factors γ_i is an iterative process because, in addition to the coefficients of variation v_{xi} of the variables, initial assumptions will have to be made regarding their mean values m_{xi}, which will have to be subsequently checked against the calculated coordinates of the design point.

Non-normal distribution of variables may be converted into the equivalent normal distribution by equalising the probability densities and the cumulative densities at the design point.

4.5   Partial safety factors in BS 5400

The British bridge code BS 5400[5] uses the Level I method (see Section 4.3) in its treatment of the two limit states of collapse and serviceability. The safety factor format used is

$$R \geqslant S$$

where:

R is the design resistance, defined as:

$\dfrac{1}{\gamma_m}$ [function of material strength f_k and geometrical parameters of the structural components]

S is the design load effect, defined as:

γ_{f3} (effects of $\gamma_{fL}\, Q_k$), where

Q_k is the nominal or characteristic loads

f_k is the characteristic strength of the material

$\gamma_{fL} = \gamma_{f1} \times \gamma_{f2}$

γ_{f1} is a partial safety factor to take account of the probability of the actual loads being higher than their characteristic values

γ_{f2} is a partial safety factor to take account of the possibility that, when various loads act simultaneously, each one of them may not be at the level of its characteristic value at the same time

$\gamma_m = \gamma_{m1} \times \gamma_{m2}$

γ_{m1} is a partial safety factor to cover the material strength in the bridge being below the characteristic value assumed in the design and specified to be achieved

γ_{m2} is the partial safety factor to cover possible reduction in the structural strength due to approximations, inaccuracies and diversities in the strength calculations, e.g. in the formulae for struts, beams, etc. and also due to geometrical imperfections

γ_{f3} is a partial factor to take account of inaccurate assessments of the effect of loading and of unforeseen stress distribution in the structure.

In BS 5400, separate values of γ_{ml} and γ_{m2} are not explicitly given; instead, a combined value of γ_m is specified. Part 2 of BS 5400[7] for loads had already specified the values of γ_{f3} and also of γ_{fL} for all the various loads acting separately or in combination with each other. The selection of appropriate γ_m values for BS 5400 Part 3[8] was done by means of a special calibration study. In this study a target reliability level was first established, which was the average reliability achieved in the past for structures, designed by commonly accepted design practices that had served well in their design lives up to the present time. Optimum sets of γ_{m1} and γ_{m2} were then derived by this study to be used in the resistance expressions in BS 5400 Part 3, so that the average reliability of the structural components thus designed for a 120-year design life was the same as the target reliability thus established, and their scatter about this average value was the minimum.

The range of structural components in bridges to be used to derive this target reliability was established from a review of the existing plate, box and truss girder bridges in Britain. From these data, the following components were selected for this calibration exercise:

(1)   compression members
(2)   tension members
(3)   beams subjected primarily to bending moments
(4)   unstiffened or vertically stiffened webs of beams subjected primarily to shear
(5)   plates in compression.

British Standard BS 153[9] was taken to represent satisfactory old design practice. BS 153 did not cover the design of continuous bridge girders, i.e. girders with high coincident bending moment and shear. There were also doubts about the adequacy of the BS 153 requirements for the design of horizontally stiffened webs and stiffened compression flanges. Hence these components were not included for deriving the target reliability.

Only dead load and vehicular live load were used in this calibration; appropriate statistical models were developed from available data on the real-life variations of these loads from the conventionally assumed design value. For dead load, the mean value was taken to be 1.05 times the design value with a coefficient of variation of 0.05 in a normal type of distribution. The statistical model for extreme single-lane vehicle loading was derived from:

(1)   an analysis of the dimensions and axle loads of vehicles passing survey stations set up on major roads, and

(2) an analysis of the frequency of occurrence of different types of vehicles passing some other survey stations.

The maximum live load on a bridge was taken to have a mean value of 1.04 times the HA loading given in BS 5400 Part 2 and a coefficient of variation of 0.09, with an extreme type 1 distribution.

The statistical model for resistance or strength of the structural components had to allow separately for the variability of the material strength and the varying degrees of approximations involved in calculating the member strengths by different mathematical formulae. An analysis of a large volume of test data on yield stress indicated a standard deviation of about 0.075, and the mean strength of the sample materials was about two standard deviations above the nominal yield stress values specified in the material standards. For calculating the failure probability of a bridge member in service, an allowance had to be made for the fact that the laboratory tests for measuring yield stress are conducted at a higher rate of straining than what a bridge structure experiences in service conditions; for this purpose the real yield stress in service loading was taken to be 15 N/mm^2 less than the measured value of test samples. Thus the actual yield stresses were taken to have a mean value and a standard deviation of 270 N/mm^2 and 20 N/mm^2 for mild steel, and 390 N/mm^2 and 25 N/mm^2 for high-yield steel.

The real strengths of the designs were predicted by using the statistical characteristics of the strength formulae proposed for the new design code. These were obtained by comparing published data on the laboratory tests of struts, ties, beams, stiffened flanges, etc.

The range of failure probabilities that were thus derived for the different structural components of past designs, covering a wide spectra of dead to live load ratios, structural geometries and materials of mild and high yield steels, are given in Table 4.2. The average value of the failure probability, after giving appropriate weighting to the various components on their usage frequencies, were calculated to be 0.632×10^{-6}, which was then taken to be the target reliability for the new design code BS 5400 Part 3.

To achieve this with the new design rules proposed, the optimum value of γ_{m1} across the whole range of components was calculated to be 1.08; and the individual optimum values for γ_{m2} for each of the components investigated, to be used in conjunction with $\gamma_{f3} = 1.1$ already specified in BS 5400 Part 2, are given in the second column of Table 4.3. The third column of this Table shows the range of failure probabilities that will result from the use of these optimised partial safety factors for the design of the various components; comparing these with the corresponding figures in Table 4.2, it may be noted that these ranges are considerably narrower than the ranges of failure probabilities achieved in old designs.

From these results it was concluded that a rationalised single γ_m

Table 4.2 Failure probabilities of old designs.

Structural components	Range of failure probability of old designs
Compression members	0.2×10^{-6} to 0.8×10^{-17}
Tension members	1.0×10^{-8} to 0.5×10^{-10}
Beam comp. flanges	0.15×10^{-11} to 0.8×10^{-19}
Beam ten. flanges	0.1×10^{-14} to 0.8×10^{-27}
Beam webs	0.5×10^{-4} to 0.3×10^{-8}
Plates in compression	0.25×10^{-5} to 0.25×10^{-9}
Weighted average	0.632×10^{-6}

($=\gamma_{m1}\gamma_{m2}$) value of 1.05 was appropriate for all components that fail by yielding in tension or compression. Slender struts exhibit a sudden drop in the load carried after reaching maximum strength, and hence a γ_m value higher than that derived as optimum was advisable; hence the same value of 1.05 was adopted for struts. The design rules for stiffened compression flanges underwent further rationalisation in the treatment of the strength and stiffness of the flange plates; to reflect the consequent changes in the mean failure probability, a γ_m value of 1.20 was finally adopted.

Buckling of compression flanges of laterally unsupported beams could not be included in the derivation of the target reliability, as the available statistical data on the modelling uncertainty were considered to be inconsistent. After further research into these data, a γ_m value of 1.20 was adopted for the proposed design rules. For buckling of webs of beams in shear, a γ_m value of 1.05 was considered appropriate for webs of low slenderness which fail primarily by yielding, while a γ_m value of about 1.25 was required for very slender webs that fail by the tension-field mechanism.

Table 4.3 γ_m factors derived from calibration.

	Optimised γ_{m2} values, for use with $\gamma_{f3} = 1.1$ and $\gamma_{m1} = 1.08$	Range of failure probabilities	Final γ_m ($=\gamma_{m1}\gamma_{m2}$) values, for use with $\gamma_{f3} = 1.1$
Struts	0.88	0.3×10^{-5} to 0.1×10^{-6}	1.05
Yielding of beam flanges	0.97	1.0×10^{-6} to 0.6×10^{-7}	1.05
Buckling of beam webs	1.12	0.25×10^{-5} to 0.2×10^{-6}	1.05 to 1.25
Buckling of stiffened flanges	1.15	1.0×10^{-6} to 0.3×10^{-6}	1.20
Buckling of plates in compression	0.97	0.15×10^{-5} to 0.15×10^{-6}	1.05
Ties	0.98	0.15×10^{-5} to 0.25×10^{-6}	1.05
Lateral buckling of beams	—	—	1.20

In the design rules, a γ_m value of 1.05 was stipulated, along with a variable adjustment factor incorporated in the strength formula. The calibration exercise showed that the use of these γ_m factors and the strength formulae given in BS 5400 Part 3, along with the γ_{fL} and γ_{f3} factors given in BS 5400 Part 2, would require approximately 6% less steel than in the previous design practice, while achieving the same degree of reliability on average. Another benefit was the very significant reduction in the scatter of the failure probability about this mean value, as can be seen from Table 4.3.

4.6 References

1. Building Code Requirements for Reinforced Concrete (ACI 318–77). American Concrete Institute, 1977.
2. Specifications for the Design, Fabrication and Erection of Steel Buildings. American Institute of Steel Construction, 1978.
3. Standard Specifications for Highway Bridges, American Association of State Highway and Transportation Officials, 1977.
4. CP 110: Part 1: 1972. The Structural Use of Concrete, British Standards Institution, London.
5. BS 5400: Part 1: 1978. Steel, Concrete and Composite Bridges. British Standards Institution, London.
6. Flint A. R. *et al.* (1981). *The Design of Steel Bridges*. Paper 2: The derivation of safety factors for design of highway bridges. Granada, London.
7. BS 5400: Part 2: 1978. Steel, Concrete and Composite Bridges: Specification for Loads. British Standards Institution, London.
8. BS 5400: Part 3: Code of Practice for Design of Steel Bridges. British Standards Institution, London.
9. BS 153: 1958: Specification for Steel Girder Bridges. British Standards Institution, London.

Chapter 5
Rolled Beam and Plate Girder Design

5.1 General features

Rolled I-section steel joists and universal beams are very convenient for bridges of up to 25 m span. Apart from a pair of vertical stiffeners over their end supports, these do not require any other fabrication. For longer spans, I-section girders made up of plates are used. Before welding became popular, flange plates were connected to a web plate by riveting through angles, as shown in Fig. 5.1(a); where a single flange plate was not adequate, several plates were used as shown in Fig. 5.1(b).

As the bending moment fell along the span, the outer plates were stopped or 'curtailed'. Welding removed the need for the flange angles and also removed the gaps between adjacent elements where water could collect and initiate rusting (see Fig. 5.1(c)). Curtailment of the flange area is achieved in welded construction by using thinner and/or narrower flange plates in regions of reduced bending moments, butt-welded to each other at the ends. There is a limit to the thickness of the flange plate that can be conveniently used, since material properties like weldability, notch toughness, through-thickness ductility and even yield stress deteriorate with increase in thickness, and risks of lamination and other inclusions increase. When a single plate is not adequate, the required flange area is provided by using several flange plates as shown in Fig. 5.1(d); the outer plates are made successively narrower than the inner ones, to which they are connected by fillet welds along the longitudinal edges. The outer plates are discontinued as the bending moments fell along the span; the discontinuity at the end of each curtailed flange plate is, however, a potential fatigue problem and needs careful detailing.

There may be other variations and combinations; for example, the flange may be made up of several plates riveted together and then welded to the web plate; the web may be made up of two thicker plates near the flanges and one thinner plate at the middle of the cross-section, butt-welded to each other along the longitudinal edges; the web depth may be varied along the span. Instead of a plate, a channel section may be used as the flange.

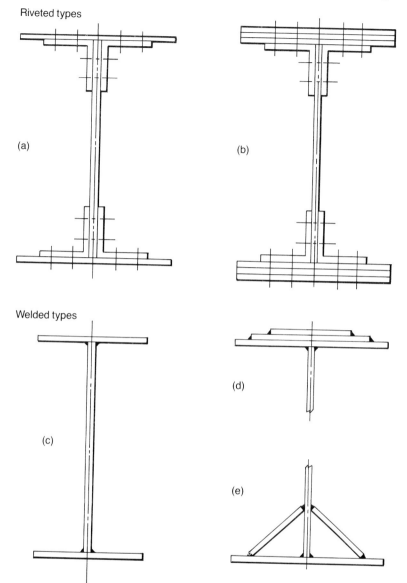

Fig. 5.1 Different types of plate girders.

Sometimes inclined plates are attached between the flange and the web, as shown in Fig. 5.1(e); they support the flange tip, as well as acting as a longitudinal stiffener to the web, and more importantly, they provide some torsional rigidity to the girder.

The material in the web is not as efficient as that in the flange in resisting bending moment. But the shear resistance of an I-beam is related only to the web area, and thus determines the minimum web area. The most efficient design of a plate girder cross-section is thus to make the web as thin and deep as possible. But deep and thin webs are liable to buckling and may require stiffening. Vertical stiffeners are provided for the webs to improve their buckling resistance, and in the case of very deep and thin webs horizontal stiffeners may also be necessary. Increased depth of beams also adds to the cost and length of the embankments in the bridge approaches.

The concrete deck on top of the top flange may be made integral with a plate girder by means of shear connectors. Plate girders are used in continuous spans by splicing adjacent lengths by either bolting or welding. Where the bridge consists of several spans, a decision has to be made whether to provide simply supported girders over each span or to make the girders continuous over several spans. The advantages of simply supported spans are:

(1)   Relative settlement of any support will not change the dead load stresses.
(2)   If each span length is within the limits of transportation, then site splicing of girders is avoided.
(3)   Expansion joints at the ends of the spans have to cope with the expansion of a single span length only.

The advantages of continuous spans are:

(1)   The number of expansion joints can be reduced, often to only one; however, the joint will have to cope with the expansion of several span lengths. Expansion joints tend to deteriorate with traffic and cause bumpiness of riding and are potential sources of trouble.
(2)   The forces due to braking and acceleration of vehicles (see Chapter 3) can be resisted at one bridge support or shared between several supports; in the case of simple spans these same forces are to be resisted fully at several supports and cannot be shared.
(3)   The number of bearings will be reduced on each pier, which can therefore be narrowed in elevation.
(4)   The structure will be generally more rigid, with reduced deflections and vibrations.
(5)   There is scope and engineering justification for increasing the girder depths over supports, thus improving the appearance of the bridge.
(6)   Finally, materials in the girder can be reduced and longer spans can be built with fewer piers, thus achieving overall economy.

There is another arrangement of suspended and cantilever spans, as shown in Fig. 1.1. This arrangement has the advantages (5) and (6) of the continuous spans, the number of expansion joints being less than for simple spans but more than for continuous spans; each support takes half the longitudinal forces and carries only one bearing, and like simple spans is unaffected by any settlement of supports. This arrangement is thus particularly suitable for bridge sites vulnerable to foundation settlements, e.g. in mining areas.

The AASHTO Specification[1] requires a minimum depth of steel beams as 1/25 of simple spans, or of the distance between points of contraflexure under dead load in the case of continuous construction. For composite beams the above limit applies to the overall depth, i.e. concrete slab plus steel girder, and there is an additional limit of 1/30 for the depth of the steel cross-section. There are also limits on the deflection under live load and impact, which often govern the depth of cross-section; these are 1/800 of the span generally, but 1/1000 of the span for bridges in urban areas used by pedestrians. Where several longitudinal girders are interconnected by crossbracings or diaphragms for efficient lateral distribution of load, the deflection for this purpose may be calculated by assuming that all the girders will deflect equally.

There are no specific limitations on girder depths or deflections in the British Standard[2], except that attention is drawn to the need for camber for the sake of appearance, drainage and headroom clearance. In the case of a nominally straight bridge, a sagging deflection exceeding 1/800 of the span is also discouraged.

5.2 Analysis for forces and moments

To design a plate girder, it is necessary to first obtain the bending moment, shear force and axial force acting on its various sections. The open cross-section of a plate girder is torsionally very flexible and hence it is generally assumed that a plate girder section cannot resist any torsion. Axial force occurs in a plate girder when the bridge deck is subjected to longitudinal forces due to, say, braking. Owing to vertical loads, axial force occurs when the plate girder is part of a portal frame or is supported by inclined cables.

If any load is applied over one side of the bridge deck, the beams directly under the load obviously deflect more than the others; the consequent transverse bending of the deck slab distributes some of the load on to beams away from the load. This transverse sharing of the load may be further improved by the provision of transverse diaphragms across the width of the bridge deck and connected to the longitudinal beams. Transverse diaphragms over the supports of the longitudinal beams prevent the latter from twisting and are virtually essential. The usefulness of intermediate diaphragms should be judged by balancing the improved lateral distribution of load against the cost of providing, connecting and maintaining them.

In a bridge deck constituted by a set of plate girders supporting a concrete deck, the most convenient way to obtain the bending moments and shear forces is by the assumption that the deck consists of a grillage of longitudinal and transverse beams. The continuous concrete slab is replaced by a series of discrete parallel beams spanning between the steel beams. If there are transverse members connected to the main longitudinal girders, then the grillage consists of these longitudinal and transverse girders. Generally the concrete deck is made to act compositely with the steel girders by the provision of shear connectors; in such cases the concrete slab is taken as a flange of the steel beam, with an effective area equal to the gross area of the slab between the steel girders divided by the modular ratio (i.e. the ratio between Young's moduli of steel and concrete). Sometimes the transverse girders are unconnected to the concrete deck; the grillage analysis in such cases must take into account two separate sets of transverse members, i.e. the concrete slab and the unconnected steel transverse girders. Shear deformation is generally ignored in grillage analysis.

Open cross-sections like I-beams have negligible torsional stiffness. The torsional stiffness of a concrete slab is also generally ignored, but may be taken into account by assuming the St Venant torsional constant as $d^3/6$ per unit width in the two orthogonal directions, when d is the slab depth.

Computer programs for grillage analysis can also easily deal with multiple-span continuous structures with constant or varying moments of inertia of the longitudinal girders. The hogging moment over intermediate supports of the longitudinal girders may cause transverse cracking of the concrete slab in these regions, thus causing a reduction in the effective moments of inertia. This can be dealt with in one of the following two ways:

(1)   A new distribution of bending moments may be determined by neglecting the concrete in the calculation of the moment of inertia of the beams over the length over supports where tensile stress in concrete was found to exceed 10% of its specified 28-day compressive strength, or over, say, 15% of the span lengths on each side of the support.

(2)   The sagging moments in the adjacent spans are increased, without reducing the hogging moment; the percentage increase is specified in BS 5400[13] as 40 times the ratio of the tensile stress in concrete to its specified 28-day compressive strength.

In the AASHTO Specification[1] empirical methods are authorised for obtaining the transverse distribution of wheel loads. For internal longitudinal beams with spacing S (metres), the bending moment may be calculated by:

(1) taking $\dfrac{S}{1.676}$ fraction of wheel loads, if $S < 4.267$ m

(2) taking the flooring to act as a simple span between longitudinal beams, if $S > 4.267$ m.

For external longitudinal beams, the flooring may be assumed to act as a simple span between longitudinal beams, except that, where there are four or more beams, the fraction of wheel loads shall not be less than:

(1) $\dfrac{S}{1.676}$, where $S \leqslant 1.829$ m

(2) $\dfrac{S}{1.219 + 0.25\, S}$, where $1.829 < S < 4.267$.

Shear forces in all beams may be calculated in the same way as bending moments, except that, for calculating end shear or reaction, the effect of a wheel load placed near that end of a beam shall always be calculated by assuming the flooring to act as a simple span. One condition for adopting this empirical method of transverse distribution is that exterior beams shall not have less carrying capacity than interior beams.

The analysis for forces and moments in individual girders is commonly based on the linear elastic theory of structural behaviour. The plastic-hinge type of analysis for continuous spans is not suitable for bridges for the following reasons:

(1) Methods currently available for analysing the lateral distribution of vehicle load over several beams are based on the principles of linear elastic behaviour of the beams and concrete slab.

(2) Moment-rotation capacity has been established only for compact beam sections.

(3) The principle of superposition does not hold good in plastic analysis and thus it will be extremely difficult to combine various live load cases and effects of temperature, etc. with the dead load.

5.3 Lateral buckling of beams

A beam required to resist the bending moment in the plane of its higher flexural rigidity may buckle out of the plane of loading, i.e. deflect laterally

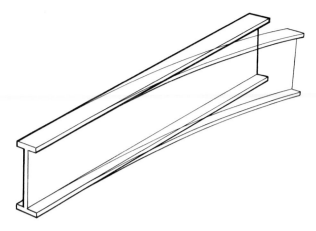

Fig. 5.2   Lateral buckling of a beam.

and twist (see Fig. 5.2) if it does not have sufficient lateral stiffness of its own or lateral support provided to it. An ideal perfectly straight beam with a high material yield stress, loaded exactly in its plane of bending containing its shear centre, will remain straight until the applied bending moment reaches a critical value M_{cr} which depends upon the length of the beam and its geometric proportions. This is the linear theory of buckling or buckling by 'bifurcation'. However, a beam with some initial misalignment and/or residual stresses and/or inclined loading will tend to deflect laterally and twist as the bending moment increases, and its failure is initiated when the in-plane bending stresses, residual stresses and stresses caused by lateral deflection and twist combine to cause yielding. This is the non-linear or 'divergence' theory of buckling. The critical bending moment of the ideal straight beam with very high yield stress will be discussed first, and then it will be described how this value is modified to take account of the onset of yielding.

5.3.1   Buckling of an ideal beam

The critical bending moment of a perfectly straight elastic beam with cross-section symmetrical about both axes is given by

$$M_{cr} = \frac{\pi}{L_e} \sqrt{\frac{EI_y \, GJ}{\alpha}} \sqrt{1 + \frac{\pi^2 \, EI_w}{L_e^2 \, GJ}} \tag{5.1}$$

where EI_y is the flexural rigidity about the minor axis
$\quad\quad\;\; GJ$ is the torsional rigidity
$\quad\quad\;\; EI_w$ is the warping rigidity

L_e is the half-wavelength of buckling, or 'effective length', as it is generally called

α is a correction factor, just less than 1.0, to correct for deflection due to bending; it is given approximately by $(I_x - I_y)/I_x$, where I_x is the major axis moment of inertia.

For the standard case of a beam of length L subjected to equal and opposite end moments, restrained at its ends against lateral deflections and twist but free to rotate in plan, and without any intermediate lateral restraint, L_e is equal to L. Equation (5.1) can also be expressed as

$$M_{cr} = \frac{\pi^2 E}{L_e^2} \sqrt{\frac{I_y \, I_w}{\alpha}} \cdot \beta \tag{5.2}$$

where β represents the contribution of the torsional rigidity of the section and is given by

$$\beta = \sqrt{1 + \frac{L_e^2 \, GJ}{\pi^2 \, EI_w}} \tag{5.3}$$

For equal flange I-sections,

$$I_w = I_y \frac{h^2}{4}$$

where h is the distance between the centroids of the flanges; hence equation (5.2) may be expressed as

$$M_{cr} = \frac{\pi^2 \, EI_y}{2L_e^2} h \frac{\beta}{\sqrt{\alpha}} \tag{5.4}$$

For the I-section the contribution of the web to the minor axis moment of inertia I_y is negligible; the critical bending moment M_{cr} can be considered to be $P_E \frac{h\beta}{\sqrt{\alpha}}$, where P_E is the Euler critical load of the compression flange. Another way of expressing equation (5.4) is

$$M_{cr} = \sigma_E \frac{Ah}{2} \frac{\beta}{\sqrt{\alpha}} \tag{5.5a}$$

or

$$\sigma_E \, Z_e \left(\frac{2A_f + A_w}{2A_f + \frac{1}{3}A_w} \right) \frac{\beta}{\sqrt{\alpha}} \tag{5.5b}$$

or

$$\sigma_E \, Z_p \left(\frac{2A_f + A_w}{2A_f + \frac{1}{2}A_w} \right) \frac{\beta}{\sqrt{\alpha}} \tag{5.5c}$$

where $\sigma_E = \dfrac{\pi^2 E}{(L_e/r_y)^2}$, i.e. the Euler stress of the whole beam

r_y = the radius of gyration of the beam about its minor axis
Z_e = elastic section modulus
Z_p = plastic section modulus
A_f = area of each flange
A_w = area of web
A = total area = $2A_f + A_w$.

Expression (5.3) for β may be simplified by making some approximations about the geometric properties of equal-flange I-sections; for example, if it is assumed that (i) the area of the web equals that of each flange, (ii) the web contribution to J is 0.6 times that of each flange, and (iii) $h = 0.95D$, D being the overall depth of the beam, then

$$I_y = \frac{1}{6} B^3 T$$

$$J = 0.87\ BT^3$$

$$I_w = I_y \frac{(0.95\ D)^2}{4}$$

$$= \frac{B^3\ D^2\ T}{26.6}$$

$$r_y^2 = \frac{B^2}{18}$$

Taking $E = 2.6G$ leads to

$$\beta = \sqrt{1 + \frac{1}{20} \left(\frac{L_e}{r_y} \frac{T}{D} \right)^2} \qquad (5.6)$$

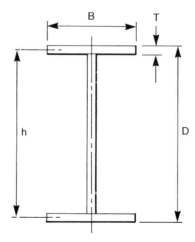

Fig. 5.3 Beam cross-section.

In the case of axially loaded struts, the most critical parameter for its strength is the slenderness ratio L_e/r. The critical, i.e. the Euler, load is given by

$$P_E = \frac{\pi^2 EA}{(L_e/r)^2}$$

and (L_e/r) can be expressed as

$$\frac{L_e}{r} = \sqrt{\frac{\pi^2 E}{\sigma_y} \frac{P_y}{P_E}}$$

where σ_y = yield stress
    A = area
    P_y = squash load = $\sigma_y A$

Similarly, for lateral buckling of beams there is a slenderness parameter λ_{LT} such that

$$M_{cr} = \frac{\pi^2 EZ_p}{(\lambda_{LT})^2} \qquad (5.7a)$$

and

$$\lambda_{LT} = \sqrt{\frac{\pi^2 E}{\sigma_y} \frac{M_p}{M_{cr}}} \qquad (5.7b)$$

where M_p is the plastic moment of resistance = $Z_p \sigma_y$.
    Putting equation (5.4) in equation (5.7b), one obtains

$$\lambda_{LT} = k\, v \left(\frac{L_e}{r_y}\right) \qquad (5.8)$$

where

$$k = \left(\frac{2A_f + \frac{1}{2}A_w}{2A_f + A_w}\right)^{\frac{1}{2}} \alpha^{\frac{1}{4}}$$

$$v = \beta^{-\frac{1}{2}} = \left(1 + \frac{L_e^2 J}{6.42 I_y h^2}\right)^{-\frac{1}{4}} \qquad (5.9a)$$

$$\approx \left[1 + \frac{1}{20}\left(\frac{L_e}{r_y}\frac{T}{D}\right)^2\right]^{-\frac{1}{4}} \qquad (5.9b)$$

    The values of k generally lie in the range 0.8 to 1.0 for rolled or fabricated I-beams with equal flanges, and near to 0.9 for universal beams. Thus a safe assumption for k is 1.0 generally and 0.9 for universal beams.
    In an I-beam with unequal flanges, the non-coincidence of the centroid and the shear centre complicates the derivation of M_{cr}, but reasonably accurate results[3] are obtained from equation (5.7a) if the following modified expression for v is adopted along with $k = 1.0$ in calculating λ_{LT} from equation (5.8):

$$v = \left[\left\{ 4N(1 - N) + \frac{1}{20}\left(\frac{L_e}{r_y}\frac{T_m}{D}\right)^2 + \psi^2 \right\}^{\frac{1}{2}} + \psi \right]^{-\frac{1}{2}} \qquad (5.10)$$

where $N = \dfrac{I_c}{I_c + I_T}$

I_c = second moment of area of compression flange about the minor axis of the beam

I_t = second moment of area of tension flange about minor axis of beam

$\psi = 0.8(2N - 1)$ for $I_c \geqslant I_t$
   $= 2N - 1$ for $I_c < I_t$

T_m = mean thickness of the two flanges.

The critical bending moment so far derived is for the case of a beam subjected to a constant bending moment along its laterally unsupported length. Other shapes of bending moment diagrams, e.g. those caused by one or more concentrated loads, or distributed loads, will be less severe, and even more so if there are hogging moments at ends. To take advantage of this, the slenderness parameter λ_{LT} may be modified by a factor η; M_{cr} then calculated on the basis of this modified value of λ_{LT} should always be taken as the critical value of the numerically maximum bending moment in the span. There is no explicit expression for η; Reference 2 gives graphs for η for distributed and concentrated loads with varying magnitudes of support moments; values of η for a few load cases are given in Table 5.1.

For beams with cross-section and bending moment falling from their maximum values along its span, λ_{LT} calculated for the cross-section subjected to maximum bending moment may be modified by a factor

$$\eta = 1.5 - 0.5\omega$$

where ω is the ratio of minimum to maximum total flange area. No further modification to λ_{LT} should be made for the varying bending moment.

When a beam twists and buckles laterally under a load applied on its top flange, then the load may either move laterally with the deflected flange, or may remain in the original plane of application. In the former case, the stability of the beam will be further reduced by the resultant torsion with respect to the supports. This effect is particularly pronounced in the case of deep and short beams[3]. A simple and conservative method of dealing with this problem is to increase λ_{LT} by 20% when the load is applied on the top flange and is free to move laterally.

So far, the slenderness parameter λ_{LT} has been obtained for the support condition of the compression flange held against lateral displacement but free to rotate in plan. If there is any restraint against rotation in plan at the supports, then λ_{LT} is reduced. Following the traditional practice for struts, a reduction factor of 0.7 may be taken when the compression flange

Table 5.1   Factor for the shape of bending moment diagram.

Shape of moment diagram	Value of η
	0.76
	0.65
	0.86
	0.98
	0.83
	0.77
	0.94
	0.62

is fully restrained against rotation in plan at both supports, and 0.85 when the restraint is partial at both supports or full at one support and no restraint at the other. A more accurate graph for the reduction factor for various degrees of rotational restraint is given in Reference 2.

In the case of cantilevers, if the support section is held against lateral displacement and twist, and the tip is free to twist and deflect laterally, then its buckled shape will be the same as that of one-half of a simply supported beam with identical support conditions; the critical value of an applied bending moment constant along the length will be given by equation (5.1), provided L_e is taken as twice the cantilever length L. For the usual cases of varying bending moments it may be noted that in a simply supported beam the cross-sections with maximum lateral displacements are subjected to the maximum bending moments, whereas in a cantilever the cross-sections with maximum lateral displacements (i.e. near the tip) are subjected to minimum bending moments. Hence the benefit from a varying bending moment diagram is much more pronounced for cantilevers. Another special feature of cantilevers in bridges is that any lateral restraint to the top flange restrains the tension flange and thus is not as effective as a restraint to the compression flange. The support condition of a cantilever may be either (i) fully fixed against rotation in both planes or (ii) continuous into an adjacent span; the latter case provides a reduced restraint against warping and can be separated into several cases of lateral restraint to the top and/or bottom flanges. The critical bending moment M_{cr} also depends significantly upon the level of application of loading, i.e. whether at top flange or from bottom flange or from the shear centre. M_{cr} for a cantilever may be obtained from equations (5.7a) and (5.8) if an appropriate effective length $L_e = K_e L$ is used, where L is the actual cantilever length and K_e is an effective length factor. The factor K_e depends not only upon all the features mentioned above, but also on the geometrical parameter given by equation (5.3). A conservative set of values for K_e is given in Table 5.2, which has been derived for the case of a single concentrated load at tip.

Table 5.2 Effective length factors K_e for cantilevers.

Support condition		Compression flange at tip laterally			
		Supported		Unsupported	
		Case 1*	Case 2**	Case 1*	Case 2**
Built-in		0.6	0.6	1.4	0.8
Continuous, with the	Supported	1.5	0.8	2.5	1.0
comp. flange laterally	Unsupported	4.5	2.4	7.5	3.0

* Case 1 is for load applied on the top flange and free to move laterally.
** Case 2 is for load applied at the level of the shear centre; these values are conservative for loading applied from the bottom flange.

5.3.2 Buckling of a real beam

As mentioned in the beginning of Section 5.3.1, the critical bending moment M_{cr} given by equation (5.1) or (5.7a) is valid for a perfectly straight beam with very high yield stress. Because of limited yield stress in real beams, there is another limit of the bending capacity given by the plastic moment of resistance M_p of the section. Real beams are likely to have some twist of the cross-section, residual stresses due to rolling or welding and possibly some inclination between the plane of bending of the cross-section and the plane of loading. Because of these imperfections, the actual moment of resistance M_R is found in laboratory tests to be lower than both M_{cr} and M_p. Figure 5.4 is a plot of these bending moments against the slenderness parameter λ_{LT}; the actual M_R is found to be distributed in the shaded area shown. This distribution is similar to that of axially loaded struts, when the actual capacity P_a, the Euler buckling load P_E and the squash load P_y are plotted against a slenderness parameter L_e/r. The Perry–Robertson formula for the axial load capacity P_a of struts can be expressed as

$$(P_E - P_a)(P_y - P_a) = \eta P_E\, P_a$$

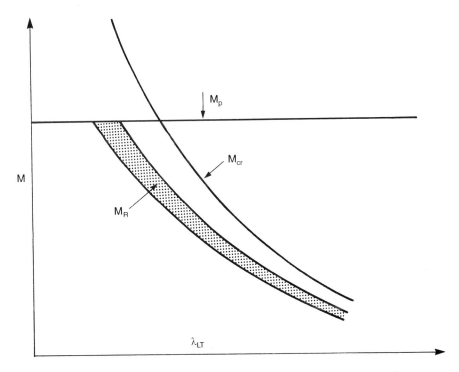

Fig. 5.4 Plot of bending moment capacities against slenderness parameter λ_{LT}.

where

$$P_E = \frac{\pi^2 EA}{(L_e/r)^2} = \text{Euler buckling load}$$
$$P_y = \sigma_y A = \text{squash load}$$
$$A = \text{area of cross-section}$$
$$r = \text{minimum radius of gyration}$$
$$\eta = \text{an imperfection parameter}$$

Similarly, the bending capacity of a beam may be expressed as

$$(M_{cr} - M_R)(M_p - M_R) = \eta M_{cr} M_R \tag{5.11}$$

where η is an imperfection parameter.

Dividing the moments by the plastic modulus Z_p, one obtains

$$(\sigma_{cr} - \sigma_b)(\sigma_y - \sigma_b) = \eta \sigma_{cr} \sigma_b \tag{5.12}$$

where $\sigma_{cr} = M_{cr}/Z_p = \pi^2 E/(\lambda_{LT})^2$ from (5.7a)

$\sigma_b = M_R/Z_p = $ a limiting bending stress.

The solution to the quadratic equation (5.12) for σ_b is given by

$$\sigma_b = \frac{1}{2}\{\sigma_y + (1+\eta)\sigma_{cr}\} - \frac{1}{2}\left[\{\sigma_y + (1+\eta)\sigma_{cr}\}^2 - 4\sigma_{cr}\sigma_y\right]^{\frac{1}{2}}$$

or

$$\frac{\sigma_b}{\sigma_y} = \frac{1}{2}\left\{1 + (1+\eta)\left(\frac{\pi^2 E}{\sigma_y\lambda_{LT}^2}\right)\right\} - \frac{1}{2}\left[\left\{1 + (1+\eta)\left(\frac{\lambda^2 E}{\sigma_y\lambda_{LT}^2}\right)\right\}^2\right.$$
$$\left. - 4\frac{\pi^2 E}{\sigma_y\lambda_{LT}^2}\right]^{\frac{1}{2}}$$
$$= \frac{1}{2}\left\{1 + (1+\eta)\frac{\pi^2}{\gamma^2}\right\} - \frac{1}{2}\left[\left\{1 + (1+\eta)\frac{\pi^2}{\gamma^2}\right\}^2 - 4\frac{\pi^2}{\gamma^2}\right]^{\frac{1}{2}} \tag{5.13}$$

when γ is a new slenderness parameter, given by $\lambda_{LT}\sqrt{\dfrac{\sigma_y}{E}}$.

Test results indicate that for small values of slenderness parameters λ_{LT} or γ, the moment of resistance M_R is not reduced below the plastic moment of resistance M_p. The imperfection parameter η is chosen in such a way as to produce (i) σ_b/σ_y equal to 1 for a stocky range (i.e. low λ_{LT} or γ values) and (ii) good correlation with test results for high values of λ_{LT} or γ.

The limiting bending stress in Reference 2 is based on the following expression for the imperfection parameter:

$$\left.\begin{array}{l}\eta = 0.12(\gamma - 1.87) \text{ for } \gamma > 1.87 \\ = 0 \text{ for } \gamma \leq 1.87\end{array}\right\} \tag{5.14}$$

This formula produces $\sigma_b = \sigma_y$ for slenderness parameter up to $\gamma = 1.87$ (or $\lambda_{LT} = 45$ for beams with yield stress $\sigma_y = 355$ N/mm^2) and also good correlation with test results for rolled and welded beam sections. Having derived the limiting bending stress from equation (5.13), the moment capacity of those beam cross-sections that are capable of developing the full plastic moment of resistance is obtained simply by multiplying σ_b by the plastic modulus Z_p. But for those beam cross-sections that are not capable of developing the full plastic moment of resistance due to the slenderness of either the flanges or the web, the use of Z_p may be unsafe when σ_b is only slightly less than σ_y; in such cases a safe and satisfactory moment capacity will be obtained by taking the lower of the two values given by (i) $\sigma_b Z_{ec}$ and (ii) $\sigma_y Z_{et}$, where Z_{ec} and Z_{et} are, respectively, the compressive and tensile elastic section moduli.

Taking equation (5.4) for the critical bending moment M_{cr}, it may be conservatively assumed that:

(1) α is equal to unity
(2)   the torsional rigidity GJ of a beam is very small compared with the warping rigidity EI_w (i.e. β is equal to unity)
(3)   the longitudinal load in the compression flange is equal to the bending moment divided by the depth h of the beam between the centroids of the two flanges.

This leads to the concept that the permissible stress in the compression flange of a beam may be taken as that of the compression flange acting as a strut. The treatment of lateral buckling of a beam in the American Code[1] for bridges is based on this concept. The critical buckling stress in the compression flange is thus taken to be its Euler buckling stress σ_E given by

$$\sigma_E = \frac{\pi^2 E}{(L_e/r_{yc})^2} \tag{5.15}$$

where r_{yc} is the radius of gyration of the compression flange about its centroidal $y-y$ axis; r_{yc} is equal to $B/\sqrt{12}$ where B is the width of the flange. The effects of imperfection and residual stresses are also treated in the manner that was originally developed by the American Column Research Council[4] for columns. In this approach, the limiting stress σ_b is related to the Euler buckling stress σ_E by

$$\sigma_b = \sigma_y\left(1 - \frac{\sigma_y}{4E}\right) \tag{5.16}$$

The above equation was originally meant to be valid in the range $\sigma_E > \sigma_y/2$. For lower values of σ_E, σ_b was taken equal to σ_E; the effects of imperfections and residual stresses in this range were covered by a slightly higher safety factor than those in the range $\sigma_E > \sigma_y/2$. In the AASHTO code, however, equation (5.16) has been adopted for lateral buckling of beams for all values of σ_E, with the same safety factor. It may be noted that equation (5.16) results in $\sigma_b = 0$ for $L_e/B = \pi\sqrt{E/3\sigma_y}$; the AASHTO code specifies an upper limit of 1.27 $\sqrt{E/\sigma_y}$ for L_e/B. This code allows the full plastic moment of resistance $(Z_p\,\sigma_y)$ to be taken when $\dfrac{L_e}{r_y} \leqslant$ $\dfrac{581}{\sqrt{\sigma_y}}$ for fairly uniform bending moment (i.e. $M_2 \geqslant 0.7M_1$), and $\dfrac{L_e}{r_y} \leqslant$ $\dfrac{996}{\sqrt{\sigma_y}}$ for varying bending moment (i.e. $M_2 < 0.7M_1$), where L_e is the distance between the effective restraints of compression flange, r_y is the radius of gyration of the whole beam about its $y-y$ axis, M_1 and M_2 are the bending moments at the ends of L_e and σ_y is in N/mm^2. The code also allows the full elastic moment of resistance $(Z_e\sigma_y)$ to be taken when

$$\frac{L_e D}{BT} < \frac{137\,900}{\sigma_y}$$

where D, B and T are the depth, flange width and flange thickness, respectively, of the beam. These limits may be compared with the limit of $\gamma = 1.87$ or $\lambda_{LT} = 874/\sqrt{\sigma_y}$, below which the British code requires no reduction in the limiting bending compressive stress due to lateral buckling.

5.4   Local buckling of plate elements

If a beam is made up of thin plate elements, i.e. thin web or flanges, then these plate elements may buckle well before the beam section reaches its overall elastic or buckling strength. Elastic buckling theories may be applied to derive the critical buckling stress of individual plate elements in the beam cross-section, i.e. the magnitude of the applied stress at which an ideal initially flat residual-stress-free plate becomes unstable and deflects out of its initially flat plane. The critical buckling stress depends upon the pattern of the applied stress, the geometry of the plate and the out-of-plane restraints on its edges. However, unlike overall buckling of beams and columns, a slender plate element may carry increased loading beyond the elastic critical value with increased out-of-plane deflection, i.e. it may have post-buckling strength.

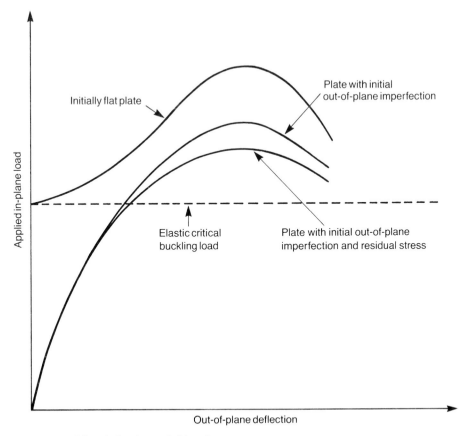

Fig. 5.5 Buckling behaviour of thin plates.

A plate with some initial out-of-flatness starts deflecting out-of-plane right from the beginning of load application, and the rate of deflection increases as the critical buckling stress is reached; in the post-buckling range the stiffness of the plate is less than that below the critical buckling level. The stiffness and strength of a plate element in the post-buckling range depend on the in-plane restraint at the edges of the plate. As a plate element starts deflecting out-of-plane, the distribution of in-plane stresses due to applied load becomes non-uniform and, in addition, bending stresses develop. As the combined in-plane and bending stresses reach the elastic limit in some parts of the plate, these parts lose their stiffness. The ultimate strength of the plate element is reached when a large part of the plate has yielded. Residual stresses in parts of the plate due to welding or rolling may bring about an earlier onset of yielding in these parts and may lower both the ultimate strength of the plate and its stiffness in the post-elastic range.

5.4.1   *Elastic critical buckling of plates*

5.4.1.1   PLATES UNDER UNIAXIAL COMPRESSION

Consider an ideally flat residual-stress-free rectangular plate simply supported along its four edges and subjected to a compressive load F per unit length uniformly distributed along two opposite edges, as shown in Fig. 5.6. At a certain value of F the flat form of equilibrium becomes unstable and the plate buckles; this instability is due to the fact that the energy of the plate in a buckled form is equal to or less than that if it remained flat under the same edge forces. The critical value of F may be determined by considering a deflected shape of the plate consistent with its boundary conditions. One such shape is given by

$$\omega = \delta \sin \frac{m\pi x}{a} \sin \frac{n\pi y}{b} \tag{5.17}$$

where ω is the out-of-plane deflection at point (x,y). In this method, the bending energy U of the plate is equated with the work done T by the applied forces due to the shortening of the plate. It can be shown[5] that

$$U = \frac{\pi^4 abD}{8} \delta^2 \left(\frac{m^2}{a^2} + \frac{n^2}{b^2} \right)^2$$

and

$$T = \frac{\pi^2 bF}{8a} \delta^2 m^2$$

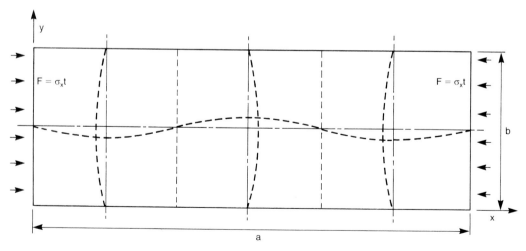

Fig. 5.6 Buckling of plates in compression.

where D is the flexural rigidity of the plate, equal to $\dfrac{Et^3}{12(1 - \mu^2)}$

> t is the thickness of the plate
> E is Young's modulus
> μ is Poisson's ratio.

The critical value of F is thus given by

$$F_{cr} = \sigma_{xcr}t = \frac{\pi^2 a^2 D}{m^2}\left(\frac{m^2}{a^2} + \frac{n^2}{b^2}\right)^2 \tag{5.18}$$

Another method for obtaining the critical value of F is to consider the St Venant differential equation of equilibrium of the plate given by

$$\frac{\partial^4 \omega}{\partial x^4} + \frac{2\partial^4 \omega}{\partial x^2 \partial y^2} + \frac{\partial^4 \omega}{\partial y^4} = -\frac{F}{D}\frac{\partial^2 \omega}{\partial x^2} \tag{5.19}$$

The deflected shape given by equation (5.17) is one solution of this differential equation. Substitution of equation (5.17) into equation (5.19) leads directly to the critical value of F given by equation (5.18). The limitation of this method is that closed-form expressions for the deflected shape of the plate are found only for a few types of applied stress patterns.

The St Venant differential equation (5.19) for a plate in compression is equivalent to the following well-known differential equation of equilibrium of a strut:

$$EI\frac{d^4 \omega}{dx^4} = -P\frac{d^2 \omega}{dx^2}$$

for which the Euler critical load P_{cr} may be obtained by substituting in the above equation a sinusoidal deflected shape.

Obviously, F_{cr} or σ_{xcr} will be minimum for $n = 1$, i.e. the plate will buckle with only one half-wave in the lateral dimension. This leads to

$$F_{cr} = \sigma_{xcr}t = \frac{\pi^2 Dk}{b^2} \tag{5.20}$$

where k is a buckling coefficient equal to $\left(\dfrac{m}{\alpha} + \dfrac{\alpha}{m}\right)^2$ where α is the aspect ratio (a/b) of the plate. For a plate of any given aspect ratio α, the number of half-waves in the direction of applied stress, i.e. in dimension a, has to be taken such as to get the minimum value of F_{cr} or σ_{xcr}. This is shown in Fig. 5.7.

The lowest value of k is 4.0, occurring when the number of half-waves equals the aspect ratio α; this value is adopted in the well-known Bryan formula for the critical buckling stress of plates in compression:

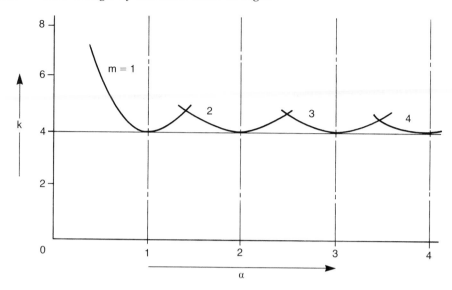

Fig. 5.7 Buckling coefficient k for plates in compression.

$$\sigma_{xcr} = \frac{4\pi^2 E}{12(1 - \mu^2)} \left(\frac{t}{b} \right)^2 \tag{5.21}$$

$$= 3.62E \ (t/b)^2$$

taking $\mu = 0.3$. It can be shown[5] that any deflected shape other than that in equation (5.17), consistent with the boundary conditions, for example, a double Fourier series, leads to F_{cr} higher than in equation (5.18).

5.4.1.2 PLATES UNDER IN-PLANE BENDING MOMENT

Consider a plate of length a, width b and thickness t, simply supported on all four edges and subjected to a linearly varying stress pattern on two opposite edges of dimension b, i.e. stresses caused by equal and opposite applied in-plane bending moments on these edges (Fig. 5.8). By taking a sufficient number of terms from a double Fourier series expression for the deflected shape of the plate, it can be shown[5] that instability occurs when the magnitude of the applied stress reaches a critical value

$$\sigma_{Bxcr} = \frac{k\pi^2 E}{12(1 - \mu^2)} \left(\frac{t}{b} \right)^2 \tag{5.22}$$

where the buckling coefficient k depends to a small extent upon the aspect ratio $\alpha = a/b$ of the plate. For $\alpha > 0.5$, the minimum and maximum values of k are 23.9 (for $\alpha = 0.67$) and 25.6 (for $\alpha = 0.5$ and 1.0); for $\alpha < 0.5$, k is larger. For the sake of simplicity k may be taken to be 24 irrespective of α, leading to

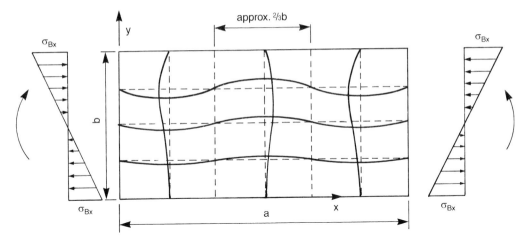

Fig. 5.8 Buckling of plate under edge bending.

$$\left.\begin{aligned}\sigma_{Bxcr} &= \frac{24\pi^2 E}{12(1-\mu^2)}\left(\frac{t}{b}\right)^2 \\ &= 21.7E\left(\frac{t}{b}\right)^2, \text{ taking } \mu = 0.3\end{aligned}\right\} \qquad (5.23)$$

5.4.1.3 PLATES SUBJECTED TO IN-PLANE SHEAR

Consider a rectangular plate with larger side a, smaller side b and thickness t, with all the edges simply supported and subjected to in-plane shear stresses τ as shown in Fig. 5.9.

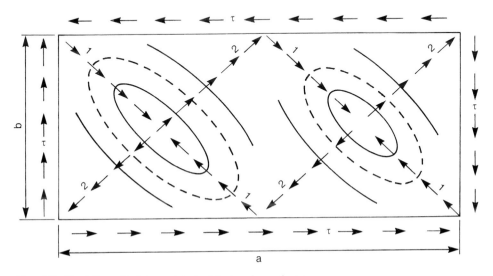

Fig. 5.9 Buckling of plates subjected to in-plane shear.

When the applied shear stress reaches a critical value the plate buckles; the buckling pattern appears in a pronounced form if there are no or little in-plane restraints on the edges. Diagonal buckles appear in elongated shapes along the direction of principal tension 1–1, i.e. several ripples forming across the direction of principal compression 2–2. Closed-form solutions of the St Venant equation are not available, but numerical solutions have been obtained[5] by the energy method by taking several terms of a Fourier series expression of the deflected form. The critical value of the shear stress can be expressed as

$$\tau_{cr} = \frac{k\pi^2 E}{12(1 - \mu^2)} \left(\frac{t}{b}\right)^2 \tag{5.24a}$$

where the buckling coefficient k is given approximately by

$$k = 5.35 + 4(b/a)^2 \tag{5.24b}$$

It should be noted that b in the above equations is always the smaller side of the plate.

5.4.1.4 PLATES SUBJECTED TO A COMBINATION OF STRESSES

A rectangular plate with simply supported edges and subjected to a combination of stresses shown in Fig. 5.10 may be assumed to reach a state of elastic critical buckling when the following condition is attained:

$$\sqrt{\left(\frac{\sigma_1}{\sigma_{1cr}}\right)^2 + \left(\frac{\sigma_2}{\sigma_{2cr}}\right)^2 + \left(\frac{\sigma_B}{\sigma_{Bcr}}\right)^2 + \left(\frac{\tau}{\tau_{cr}}\right)^2} = 1 \tag{5.25}$$

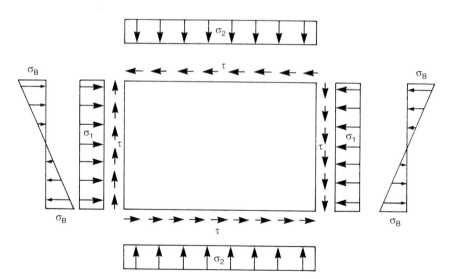

Fig. 5.10 Plate subjected to a combination of stresses.

In the above equation σ_1, σ_2, σ_B and τ are the individual stress components and σ_{1cr}, σ_{2cr}, σ_{Bcr} and τ_{cr} are the magnitudes of these individual stress components that acting alone on the plate will cause elastic critical buckling; the values of σ_{1cr} etc. have been derived in the preceding Sections for various plate aspect ratios $\propto = a/b$ and slenderness ratios b/t. Equation (5.25) is an approximate, lower-bound, simple and umbrella-type relationship that covers reasonably satisfactorily theoretical solutions for many specific stress patterns and plate geometries obtained by research[5, 6] using various theoretical techniques. It may be noted that the direction of τ and σ_B is immaterial for plate buckling and this is also reflected in equation (5.25); this equation has also been found to be reasonably accurate for negative values of σ_x or σ_y, i.e. tensile stresses, provided they are numerically not greater than about $0.5\,\sigma_{xcr}$ or $0.5\,\sigma_{ycr}$, respectively.

5.4.1.5 PLATES WITH EDGES CLAMPED

In the preceding cases, the four edges of the rectangular plates have been assumed to be hinged. Another ideal edge condition is full restraint against any rotation, i.e. fully clamped. This edge condition increases the elastic critical buckling stresses above the values for hinged edges, for example the minimum buckling coefficients for the two cases of longitudinal compression and pure bending increase from 4.0 and 23.9 mentioned in earlier Sections to 6.97 and 39.6, respectively. However, this ideal condition of full fixity against any edge rotation is difficult to achieve in real structures; very substantial structural members will have to be attached on the edges to achieve any substantial degree of fixity. Secondly, the increase in the elastic critical buckling stress attained by edge fixity is not accompanied by a similar percentage increase in the ultimate strength of the plate; in many laboratory tests, the ultimate strength has been found to be very little improved by the edge fixity. For these reasons, the design of plated structures on the basis of ultimate limit strength is normally based on the assumption that the supported edges of the plates are not restrained against rotation.

5.4.2 *Post-buckling behaviour of plates*

In the preceding Section the magnitude of the applied in-plane stress at which an initially flat plate first buckles has been derived for various stress patterns. Depending on the in-plane restraints on its edges, a buckled plate can carry stresses higher than this elastic critical stress, with the buckles growing in size but still in a stable condition. If such stable buckles are acceptable, then considerable gain in the strength of such plates is thus possible.

If the transverse edges of a rectangular initially flat plate approach each other by a uniform amount across the width of the plate, then longitudinal compressive stresses will also be uniform across the width, until the elastic critical stress is reached and the plate buckles. After buckling, however, the condition changes. Imagine the plate to be made up of a number of longitudinal strips; the total distance by which the extremities of each longitudinal strip will approach each other will be the sum of

(1)   the axial shortening of the strip due to the longitudinal compressive stress carried by it, and
(2)   the reduction in the chord length due to the bowing out-of-plane, or buckling, of the strip.

It has been shown in the preceding Section that a rectangular plate under longitudinal compression buckles with only one half-wave across its width; hence the longitudinal strip along the longitudinal centre line of the plate will bow out-of-plane, or buckle, by a bigger amount than the strips nearer the longitudinal edges. If, after the onset of buckling, the transverse edges of the buckled plate continue to approach each other by a uniform amount across the width, it follows therefore that a central longitudinal strip will undergo less axial shortening and consequently carry lower longitudinal compressive stress, than the strips nearer the longitudinal edges; this redistribution of the longitudinal stress, i.e. a transfer of stress from the relatively flexible central region to the two regions near the longitudinal edges, is shown in Fig. 5.11.

The conditions of in-plane restraint in the transverse direction along the two longitudinal edges of the plate influence its post-buckling behaviour and stress distribution. If these edges are free to move in-plane in the transverse direction, then stresses in the transverse direction are zero along these edges and are also small in the interior of the plate; the longitudinal edges will, however, not remain straight but will pull in more in the crest and trough regions of the buckles and less near the nodal lines. If the longitudinal edges are prevented against in-plane movement in the transverse direction, then significant transverse tensile stresses develop; these will be higher in the crest and trough regions of the buckles and less near the nodal lines. An intermediate state of transverse restraint along the longitudinal edges is when the edges are constrained to remain straight though allowed to pull in, the net of average transverse stress along the edge being zero; in this case the transverse tensile stresses in the crest/ trough regions are balanced by transverse compressive stresses in the regions adjacent to the nodal lines. These transverse stress distributions are shown in Fig. 5.12. Transverse tensile stresses provide some stability against buckling and reduce the out-of-plane deflection. In the post-buckling

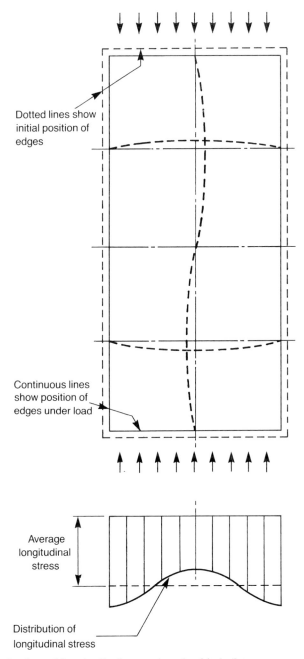

Fig. 5.11   Distribution of longitudinal stress in a buckled plate.

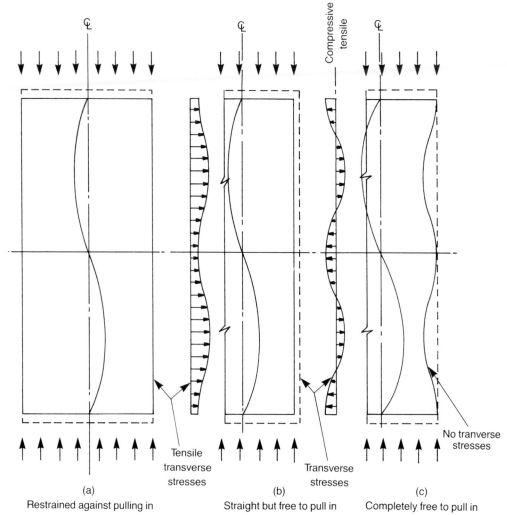

Fig. 5.12(a)   Transverse stresses in buckled plates, with different longitudinal edge
          conditions.

stage, the axial stiffness of the plate, i.e. the longitudinal compressive
force per unit area of cross-section divided by the longitudinal shortening
per unit length, is given by:

(1) $0.75E$, when the longitudinal edges are restrained against any pulling
      in
(2) $0.5E$, when the longitudinal edges are constrained to remain straight
      but free to pull in
(3) $0.41E$, when the longitudinal edges are completely free to pull in.

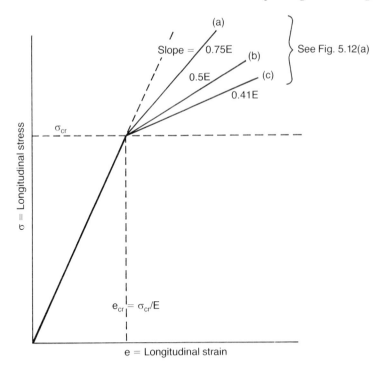

Fig. 5.12(b)   Post-buckling stiffness of plates, with different longitudinal edge conditions.

As the buckling of the plate continues with increase in applied loading, the maximum longitudinal stress along the longitudinal edges reaches yield stress, or alternatively the longitudinal stress along an interior longitudinal strip combined with flexural stresses in the plate due to the buckles reaches the yield stress locally on the surface. In the former case that stage signifies the ultimate state of the plate; in the latter case, further increase in applied loading spreads the yielding gradually over a wider surface area and through the thickness of the plate, though with a gradually falling plate stiffness and till no increase in load can be resisted.

In a plate subjected to shear stresses there is often a substantial reserve of strength after the elastic critical buckling value of the shear stress is reached. In a state of pure shear stress there are principal tensile and compressive stresses in directions at $45°$ to the direction of the shear stress, as shown in Fig. 5.13. Buckling of the plate is caused by the principal compressive stress in direction 2–2, resulting in ripples forming with their crests stretched in the direction of principal tension, i.e. 1–1. Because of the ripples, the compressive stress cannot increase beyond the value at the critical buckling stage, but the diagonal tension continues to increase with applied shear. The increased diagonal tensile stresses form what is known as

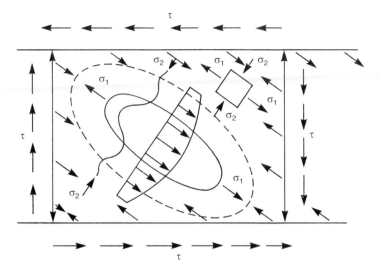

Fig. 5.13   Tension field due to shear stress.

a tension field. These tensile stresses have to be resisted on the horizontal and vertical boundaries. The flexural rigidity of the flanges resists the pulling-in effect of the tension field, while the transverse web stiffeners act as struts to provide support to the flanges, thus forming a truss-type system of forces. The ultimate shear capacity is reached when the plate in the diagonal tensile band yields and also plastic hinges are formed in the boundary members due to the pulling-in forces in the plate. In the post-buckling stage, the stiffness of the plate against shearing deformation ranges from 0.75 to 0.90 times the shear modulus G, depending upon the flexural stiffness of the boundary members. In a vertically stiffened web plate of a girder, the pulling-in forces on the two sides of intermediate vertical stiffeners balance, leaving only vertical compression to be resisted by them. But the flexural stiffness of the flanges has a significant influence on the magnitude and pattern of the tension field.

A plate subjected to pure bending stresses also has significant post-buckling capacity. As the applied stresses reach the elastic critical value, buckles appear in the compressive part of the plate. With further increase in loading, the distribution of the bending stresses changes, with no further compressive stress in the buckled portion, but the rest of the plate continues to resist the increase in loading, as shown in Fig. 5.14.

5.4.3   *Effect of residual stresses*

Residual stresses in rolled steel sections are mainly caused by uneven cooling after rolling; in sections fabricated by welding together several

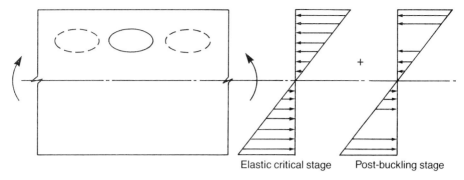

Elastic critical stage   Post-buckling stage

Fig. 5.14 Post-buckling behaviour under bending stress.

plates, the residual stresses are caused by the shrinkage of the material in and adjacent to the weld. In rolled I-sections, the flange tips cool first, but the delayed cooling of the interior parts causes compressive stresses along the flange tips; the junction between the flange and the web stays hot the longest and is thus subjected to tensile stresses as the adjacent colder parts tend to prevent its shrinkage. For equilibrium, the tensile and compressive longitudinal forces in the cross-section must balance. A typical residual stress pattern in a rolled I-section is shown in Fig. 5.15. Compressive stress along the tip may be of the order of $100-150$ N/mm^2.

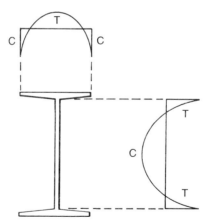

Fig. 5.15 Residual stresses in rolled sections.

Welding or flame-cutting is associated with very high temperatures in a localised strip. Shrinkage due to cooling of this strip is resisted by the remaining cold portion of the steelwork. As a result, the strip adjacent to the weld or flame cut is subjected to high tensile strains which may be several times the yield strain, and the rest of the steelwork is subjected to compression. A typical pattern is shown in Fig. 5.16. The shrinkage force

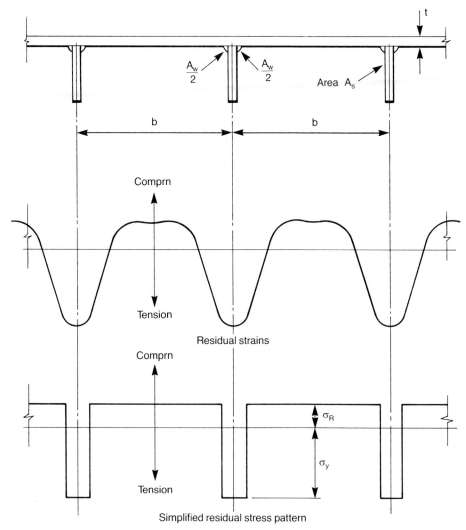

Fig. 5.16   Residual strains and stresses due to welding.

due to welding can be expressed as (CA_w), where A_w is the cross-sectional area of the weld deposited and C is a constant dependent upon the welding process adopted. C has been found experimentally to vary from 7.5 to 12.5 kN/mm^2; the lower values in this range are typical of manual welding and the higher values are associated with submerged arc welding. In multi-welds, if the steelwork is allowed to cool down to room temperature between successive weld passes, A_w is the area of weld deposited in one pass. A simplified pattern of the residual stresses may be derived by assuming the tensile stresses in the strip adjacent to the weld and the compressive stresses in the remaining area to be uniform in their respective areas, and the former to be equal to the yield stress σ_y of the steel. The cross-sectional area of the strip adjacent to the weld will then be given by (CA_w/σ_y). The compressive stress σ_R in the remaining area will have to balance the tensile force in the yielded area and thus will be given by

$$\sigma_R = \frac{CA_w}{A_g - (CA_w/\sigma y)}$$

where A_g is the total cross-sectional area of the welded assembly. In Fig. 5.16, A_g is equal to $(A_S + bt)$ per panel.

Taking an example of a 15 mm plate of yield stress 355 N/mm^2 stiffened by stiffeners of size 200×20 at 300 mm centres, each of which is welded to the plate by two 6 mm size manual fillet welds,

$A_w = 40$ mm^2, allowing for a small concavity of the weld surface
$CA_w = 7.5 \times 40 = 300$ kN
$A_g = 300 \times 15 + 200 \times 20 = 8500$ mm^2

$$\sigma_r = \frac{300\,000}{8500 - \left(\dfrac{300\,000}{355}\right)} = 39.2 \text{ N/mm}^2$$

As compressive and tensile residual stresses in the cross-section balance, residual stresses do not cause any resultant axial force or bending moment on any cross-section. If it is possible for the whole of the cross-section to squash, i.e. yield in compression, then the axial load or bending moment capacity of the cross-section will be unaffected by residual stresses. However, those parts of the cross-section where the residual stress is of the same nature as the applied stress will reach yield stress earlier. For example, a welded box column that can resist 300 N/mm^2 applied compressive stress in a residual-stress-free condition may have some parts of its walls yielding at an applied stress level of 260 N/mm^2 if the welding compressive residual stress in the zone is 40 N/mm^2. For further applied loading these initially yielded parts of the cross-section will not contribute any resistance and thus the effective stiffness of the cross-section will fall. In the case of slender struts, the flexural rigidity of the cross-section is the most important parameter for their strength, and this is the reason why residual stresses cause a reduction in their strength. The same applies to that type of welded stiffened plate assembly where the buckling of the stiffener out of the plane of the plate is more critical than the buckling of the plate between the stiffeners.

For the strength of plate panels in compression the important question is whether the plate panel, in a residual-stress-free condition, behaves in a

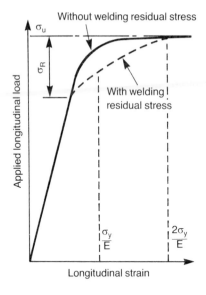

(a) Plate with stable post-buckling behaviour

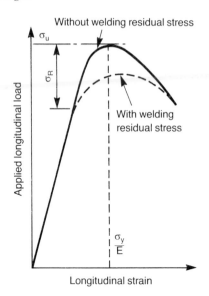

(b) Plate with unstable post-buckling behaviour

Fig. 5.17   Behaviour of plates in compression, with and without welding residual stress.

stable or unstable manner when the axial straining is continued after it reached its maximum capacity. This is illustrated in Fig. 5.17.

In the case of plate (a), in the residual-stress-free condition the plate continues to sustain the peak load σ_u when further strained; the same plate cross-section with welding residual compressive stress σ_R departs from its initial primarily linear behaviour at an applied stress $(\sigma_u - \sigma_R)$, but with further applied strains reaches an ultimate load σ'_u which is almost the same as or just below σ_u. In the case of plate (b), in the residual-stress-free condition the plate sheds off its axial load when the straining is continued beyond the peak strength σ_u; with residual stress the plate reaches a peak strength σ'_u which is substantially below σ_u. Stocky plates, i.e. with b/t ratios such that the critical buckling stress σ_{cr} is more than twice the yield stress σ_y, behave like Fig. 5.17(a); so do very slender plates, i.e. those with critical buckling stress less than half the yield stress. But plates with intermediate slenderness, i.e. with critical buckling stress between half and twice the yield stress, tend to behave as in Fig. 5.17(b), and their strength is thus affected by the level of welding compressive residual stress.

Welding residual stresses are less important for plates subjected to in-plane shear or bending stress than plates under longitudinal compression. This is because the applied stresses and the welding residual stresses are likely to be of different natures in different parts of the plates.

5.4.4 *Effects of initial out-of-plane imperfections*

Sections 5.4.1 and 5.4.2 dealt with the buckling behaviour of ideally flat plate panels, i.e. plates without any initial out-of-plane deviations. Plates in real-life fabricated structures are likely to have some initial out-of-plane deviations. Instead of the ideal bilinear behaviour shown in Fig. 5.12(b), the real behaviour of such plates will be as shown in graph (b) of Fig. 5.18. The initial out-of-plane deviation will start growing in depth from the beginning of loading. There will not occur any sudden change in the rate of in-plane or out-of-plane deformation at the theoretical elastic critical buckling stress, but the rate will gradually increase till the maximum resistance or ultimate strength of the plate is reached. Thereafter as the resistance falls, the in-plane and out-of-plane deformations continue to grow.

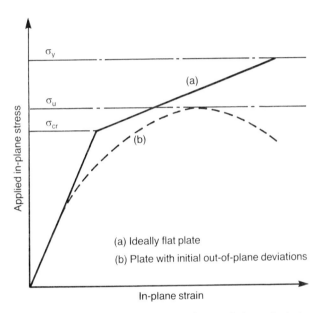

Fig. 5.18 Buckling behaviour of plates with initial out-of-plane deviations.

The in-plane edge conditions of the plate influence its behaviour right from the beginning. As in the case of perfectly flat plates, the ultimate strength is reached by a combination of redistribution of the in-plane applied stresses and the growth of bending stresses due to the out-of-plane deviations. The quantitative effects of initial out-of-plane deviations depend on:

(1) the magnitude of the deviations

(2)   the pattern of the deviations
(3)   the type of the applied stresses on the plate
(4)   the in-plane and out-of-plane edge conditions of the plate
(5)   the slenderness (i.e. the width to thickness ratio) of the plate and to a
      lesser extent its aspect ratio (i.e. the ratio of length to width).

Obviously, the larger the deviations, the worse are the effects. Initial
deviations in the pattern of the elastic critical buckling mode of the plate
have the worst effects. For example, in a plate with an aspect ratio of 3,
subjected to compression on the shorter edges, the worst pattern of initial
deviations will be the one with one sine wave across the width and three
sine waves in the form of ripples along the length. Plates subjected to in-
plane compression along one or both directions are most affected by initial
deviations; plates in shear are least affected. In-plane restraints on the
edges reduce the effects of initial deviations. But the beneficial effects of
out-of-plane edge fixities (i.e. clamped against rotation) are less pronounced in the case of initially curved plates than the improvement in the
elastic critical buckling stress of a perfectly flat plate. Plates with low
slenderness, e.g. the width–thickness ratio less than, say, 20 in the case
of longitudinal in-plane compression or 50 in the case of in-plane bending
or shear, are able to reach their squash loading, i.e. applied stress can
be as high as the yield stress, in spite of any out-of-plane deviations. A
more accurate measure of the plate slenderness is the parameter $\dfrac{b}{t}\sqrt{\dfrac{\sigma_y}{E}}$,

which indicates that plates with identical dimensions but of higher yield
stresses are effectively more slender, in the sense that the ratio of their
ultimate strength to yield strength is more reduced.

The effects of initial imperfections and residual stress on the strength of
plated structures were highlighted by the Merrison Inquiry[7] into the
failure of several box girder bridges in the early 1970s. Extensive theoretical
and experimental investigations also took place at Imperial College,
London, and elsewhere. The theoretical methods were based on the elastic
large-deflection equations first suggested by Von Karman for describing
the buckling behaviour of plates. Non-linearity in the material behaviour
during/after yielding was dealt with by adopting:

(1)   an ideal elastic/perfectly plastic behaviour, i.e. Hooke's law of pro-
      portionality between stress and strain up to yielding, and no strain-
      hardening in the post-yielding stage
(2)   a criterion for stresses to cause yielding; Hencky–Mises' criterion is
      used for this purpose, according to which yielding occurs when an
      equivalent stress σ_e reaches the yield stress σ_y of the material, σ_e

being given by the following formula for a two-dimensional stress field:

$$\sigma_e = [\sigma_1^2 + \sigma_2^2 - \sigma_1\sigma_2 + 3\tau^2]^{\frac{1}{2}}$$

(3) a relationship between stresses and strains during yielding; the flow rules due to Prandtl–Reuss are used for this purpose, which are based on the two assumptions that no permanent change of volume occurs and the rate of change of plastic strain is proportional to the derivatives

$$\partial\sigma_e/\partial\sigma_1, \ \partial\sigma_e/\partial\sigma_2 \text{ and } \partial\sigma_e/\partial\tau.$$

The magnitude of the initial out-of-plane imperfections in the plates was quantified from physical surveys of levels and patterns of imperfections in real structures, and was also related to the construction tolerance specified in the specification for steelwork construction. Thus the fabrication tolerance Δ is given by

$$\Delta = \frac{G}{250} \sqrt{\frac{\sigma_y}{245}}$$

where G is the gauge length over which the geometric imperfection was measured and is given by twice the smaller dimension b of the plate. The initial inperfection assumed for the design strength of the plate is 1.25 Δ and is thus given by

$$\delta = \frac{b}{200} \sqrt{\frac{\sigma_y}{245}}$$

σ_y being the specified yield stress of the plate in N/mm². The pattern of the initial inperfection is assumed to be sinusoidal in both directions, and the number of half-waves in each direction was varied to obtain the worst results for the stress pattern under consideration. Generally this coincided with the elastic critical buckling mode for the stress pattern. The main reason for relating the imperfection to the yield stress σ_y was that the non-dimensional imperfection parameter δ/t could be related to the non-dimensional slenderness parameter $\dfrac{b}{t}\sqrt{\dfrac{\sigma_y}{E}}$ as follows:

$$\frac{\delta}{t} = 0.145 \frac{b}{t} \sqrt{\frac{\sigma_y}{E}}$$

and this allowed the non-dimensional strength parameters σ_u/σ_y to be obtained from the same strength graphs for all strength grades of steel. It was found that plates subjected to in-plane shear or bending pattern of stresses were not very sensitive to initial out-of-plane deviations, but plates subjected to in-plane compression were, and especially in the slenderness range of $40 < \dfrac{b}{t}\sqrt{\dfrac{\sigma_y}{245}} < 60$. For example, in the above slenderness range, doubling the initial imperfection amplitude from $\dfrac{b}{400}$ to $\dfrac{b}{200}$ or from $\dfrac{b}{200}$ to $\dfrac{b}{100}$ reduced the strength of the plate by up to 10%.

Regarding initial welding residual stresses, it was found that both plates under compression and plates under shear were sensitive, particularly in the slenderness range $\dfrac{b}{t}\sqrt{\dfrac{\sigma_y}{245}}$ of 40 to 60 and 90 to 150, respectively, though by a lesser degree in the case of shear. For example, welding residual compressive stress equal to 10% of the yield stress caused up to 10% reduction in the compressive strength of a plate with initial imperfection amplitude of $b/200$, and an increase of welding compressive stress to 33% of the yield stress caused another 10% reduction in the strength of that plate. A plate of slenderness $\dfrac{b}{t}\sqrt{\dfrac{\sigma_y}{245}} = 120$, and initial imperfection amplitude $\dfrac{b}{200}$ had its shear strength reduced by 5% by welding compressive stress equal to 10% of the yield stress. Thus, even for plates that were sensitive to initial geometric imperfections and welding residual stresses, assumptions of $\dfrac{b}{200}\sqrt{\dfrac{\sigma_y}{245}}$ and $0.1\,\sigma_y$ for the two parameters yielded plate strengths that were not substantially decreased with further increase in these two parameters. Hence for design strength, the level of welding residual compressive stress was assumed to be 10% of the yield stress.

In these large-deflection elasto-plastic computer analyses, all four edges were taken as simply supported, i.e. no restraint against rotation. The in-plane loading was applied as linear displacements of the edges, the resultant of the edge stresses being taken as the applied edge load. In the case of pure in-plane bending applied on the transverse edges, the longitudinal edges were held in the shape of the simple bending curvature instead of being held straight.

Plate strength curves are given in the British Standard BS 5400: Part 3 for the two cases of

(1)   no in-plane restraint at the longitudinal edges, and
(2)   longitudinal edges held straight (or in a pre-determined configuration, e.g. for pure bending case), but allowed to move inwards in-plane

for the three applied stress patterns of (i) uniaxial longitudinal compression, (ii) in-plane shear, and (iii) in-plane pure bending. For a combination of these stresses, the interaction formula of equation (5.25) for elastic critical buckling was found to produce results that were similar to those obtained from the large-deflection elasto-plastic computer analysis. In addition, a yielding check is specified using the Hencky–Mises equation of

$$\{\sigma_1^2 + \sigma_2^2 - \sigma_1\sigma_2 + 3\tau^2\}^{\frac{1}{2}} \not> \sigma_y$$

where σ_1 and σ_2 are stresses in two orthogonal directions and τ is the shear stress.

5.4.5 Tension field in girder webs

The preceding Section dealt with the inelastic buckling strength of plate panels with two types of in-plane restraints on edges. In the case of slender webs of plate girders with transverse (i.e. vertical) stiffeners, the flanges usually constitute an intermediate level of restraint between the two extremes of the longitudinal edges being either fully free to pull in or fully held to remain straight. The shear buckling strength of such webs can be obtained by an analysis of the 'tension field' described in Section 5.4.2 and represented pictorially in Fig. 5.13. As can be seen in this diagram, the diagonal stresses in the web tend to pull the flanges inwards in the plane of the web. The relative flexural rigidity of the flanges determines the amount they are pulled in, which in turn determines the spread or width of the tension field. For example, with very rigid flanges the tension field spreads uniformly over the whole panel; but with very flexible flanges, the tension field band is narrow and is anchored primarily from only the corners of the panel.

The tension-field mechanism was first identified in plate girder railway bridges towards the end of the last century. In the 1930s Wagner developed a diagonal-tension theory for the web strength in aircraft fuselages on the assumption of negligible flexural rigidity of the web plate (i.e. the elastic critical buckling strength equal to zero) and infinite flexural rigidity of the flanges in the plane of the web. According to this theory the ultimate strength of the web in pure shear is given by $\sigma_y/\sqrt{2}$ and is reached when the diagonal tensile stress at 45° inclination to the horizontal boundaries everywhere in the web panels attains the yield stress of the web material. Since then, many tension-field models have been proposed by theoretical and experimental researchers. They are all based on the observation in tests that the tensile membrane stress in the web, combining with the shear stress that is present prior to buckling, causes yielding of the web, and subsequent failure of the whole panel is brought about by the formation of a mechanism involving the yielded zone of the web and plastic hinges in the flange.

Basler and Thurlimann were the first to develop a tension-field model for plate girders. They ignored the flexural rigidity of the flanges and assumed an off-diagonal tension field band shown in Fig. 5.19, with its inclination θ_t such as to produce the maximum shear strength. The shear strength τ_u was deduced to be

$$\tau_u = \tau_{cr} + \tfrac{1}{2} \sigma_t \sin \theta_d \qquad (5.26)$$

where τ_{cr} = elastic critical buckling stress in shear
σ_t = membrane tensile stress in the tension field
θ_d = angle of the panel diagonal with the horizontal.

The magnitude of the membrane tensile stress σ_t was taken to be such that, when combined with the elastic critical shear stress τ_{cr}, yielding occurs in the tension field band as per the Hencky–Mises criterion. Thus

$$\sigma_t = [\sigma_{yw}^2 + \tau_{cr}^2 \, (\tfrac{9}{4} \sin^2 2\theta_t - 3)]^{\frac{1}{2}} - \tfrac{3}{2}\tau_{cr} \sin 2\theta_t \qquad (5.27)$$

where σ_{yw} is the yield stress of the web material. Basler also gave an approximate and conservative expression for σ_t as

$$\sigma_t = \sigma_{yw} - \tau_{cr} \, \sigma_{yw}/\tau_{yw} \qquad (5.28)$$

and substituting this in equation (5.26) leads to

$$\tau_u = \tau_{cr} + \frac{1}{2}\sigma_{yw}\left(1 - \frac{\tau_{cr}}{\tau_{yw}}\right) \sin \theta_d \qquad (5.29)$$

It was pointed out later that equation (5.28) actually gave the shear strength of a panel with a tension field spread over its entire area and thus overestimated the theoretical shear strength of a girder with flanges incapable of supporting the in-plane pulling forces. The correct solution of the Basler model of negligible flange rigidity is given by

$$\tau_u = \tau_{cr} + \sigma_{yw}\left(1 - \frac{\tau_{cr}}{\tau_{yw}}\right) \frac{\sin \theta_d}{2 + \cos \theta_d} \qquad (5.30)$$

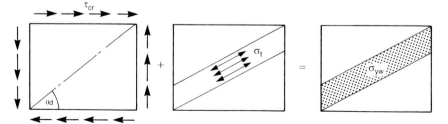

Fig. 5.19 Tension field of Basler and Thurlimann.

Many tension-field mechanisms have been postulated on the assumption that the ultimate shear capacity is the sum of the resistance of the following three separate and successively occurring mechanisms, as shown schematically in Fig. 5.20: (a) a pure shear field, (b) a diagonal tension field, and (c) a frame mechanism, involving the flanges.

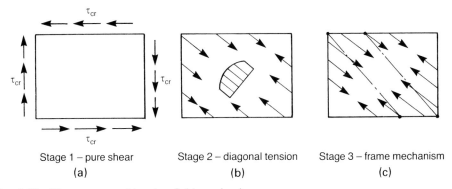

Stage 1 – pure shear   Stage 2 – diagonal tension   Stage 3 – frame mechanism

(a)   (b)   (c)

Fig. 5.20 Three stages of tension-field mechanism.

Mention may be made of the mechanisms proposed by Fujii and also by Ostapenko and Chern, in which there is an off-diagonal central band of yielding where the magnitude of the membrane tension takes account of the stress that existed at the onset of buckling, and smaller tension in the two triangular areas that can be resisted by the flanges — see Fig. 5.21.

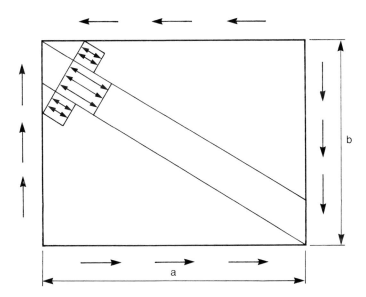

Fig. 5.21 Tension-field mechanisms of Fugi and Ostapenko/Chern.

Ostapenko/Chern gave the following expression for the ultimate shear strength:

$$V_u = \tau_{cr}\, bt + \frac{1}{2}\sigma_t\, bt\, [\sin 2\theta_t - (1 - p)\,\frac{a}{b} + (1 - p)\,\frac{a}{b}\cos 2\theta_t]$$
$$+ \frac{2}{a}\, [m_{pb} + m_{pt}] \tag{5.31}$$

where p is the ratio of tensions in the outer and inner bands, m_{pb} and m_{pt} are the plastic moments of resistance of the bottom and top flanges, respectively, σ_t is the tension in the central band, given by equation (5.27). V_u is differentiated with respect to θ_t to find the maximum V_u (to avoid this the authors gave some formulae for the shear strength). τ_{cr} is modified to take account of three levels of buckling, namely (i) the unmodified elastic critical buckling value, with the web panel edges rotationally fully restrained by the flanges but simply supported by the transverse stiffeners, (ii) modified for inelastic buckling if the elastic critical buckling stress exceeds half the shear yield stress, (iii) modified for strain hardening if it exceeds the shear yield stress.

A versatile model that has been used in Reference 2 is due to Rockey, Evans & Porter[8], with some minor modifications. The limit of the first stage of pure shear stress is assumed to occur when the elastic critical value τ_{cr} is reached. However, to allow for initial imperfections and residual stresses, a limiting value τ_l is taken less than τ_{cr} when τ_{cr} is greater than 0.8 times the shear yield stress τ_y, and equal to τ_y when τ_{cr} is greater than 1.5 τ_y. The exact values of τ_l are given by

$$\frac{\tau_l}{\tau_y} = \frac{904}{\beta^2}\,,\text{ when } \beta \geqslant 33.62$$
$$= 1.0,\text{ when } \beta \leqslant 24.55$$
$$= (1.54 - 0.22\beta),\text{ when } 24.55 < \beta < 33.62$$

where

$$\beta = \frac{1}{\sqrt{K}}\frac{b}{t_w}\sqrt{\frac{\sigma_{yw}}{355}}$$

$$k = 5.34 + \frac{4}{\phi^2},\text{ when } \phi \geqslant 1$$

$$= \frac{5.34}{\phi^2} + 4,\text{ when } \phi < 1$$

$$\phi = \text{aspect ratio } \frac{a}{b}$$

In the second stage, tensile membrane stresses τ_t develop in the web panel in a direction which does not necessarily coincide with the diagonal. The maximum shear capacity is reached in stage three when (i) the pure shear stress τ_l of the first stage and the membrane tensile stress τ_t of the second stage cause yielding in accordance with the Hencky-Mises yield criterion (i.e. equation (5.27) with τ_{cr} replaced by τ_l given above), and (ii) plastic hinges occur in the flanges, which together with the yielded zone WXYZ form a plastic mechanism.

Considering the virtual work done in this mechanism and adding the resistance of the three stages, one obtains the ultimate shear capacity as

$$V_u = \frac{4M_P}{c} + c\, t_w\, \sigma_t\, \sin^2 \theta_t + \sigma_t\, b\, t_w\, (\cot \theta_t - \phi) \sin^2 \theta_t + \tau_l\, b\, t_w$$

(5.32)

where M_p is the plastic moment of resistance of the flange, c is the distance of the internal plastic hinge (see Fig. 5.22) in one flange from the corner, and σ_t is the membrane tensile stress, given by equation (5.27), with τ_l replacing τ_{cr}.

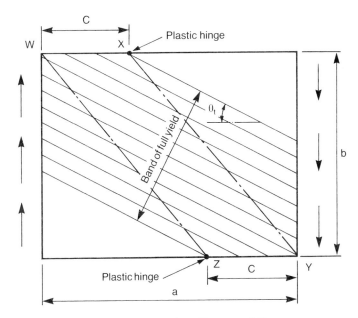

Fig. 5.22   Tension-field mechanism of Rockey, Evans and Porter.

The equilibrium condition of the flange between W and X (or between Z and Y) in Fig. 5.22 leads to

$$c = \frac{2}{\sin \theta_t} \sqrt{\frac{M_p}{\sigma_t t_w}}, \text{ but } \not> a$$

Putting this expression for c in equation (5.32) leads to the ultimate shear capacity τ_u being given by

$$\frac{\tau_u}{\tau_y} = \frac{\tau_l}{\tau_y} + 5.264 \sin \theta_t \sqrt{m} \sqrt{\frac{\sigma_t}{\tau_y}} + (\cot \theta_t - \phi) \sin^2 \theta_t \frac{\sigma_t}{\tau_y}$$

$$(5.33)$$

where

$$m = \frac{M_p}{b^2 t_w \sigma_{yw}}$$

$$\tau_y = \frac{\sigma_{yw}}{\sqrt{3}}$$

σ_{yw} being the yield stress of the web.

Different values of the inclination θ_t of the membrane tension are tried to get the highest value of τ_u. It has been found by parametric studies that θ_t is never less than $\frac{1}{3} \cot^{-1} \phi$, nor ever more than $\frac{\pi}{4}$ or $\frac{4}{3} \cot^{-1} \phi$. For calculating M_p, the flange is assumed to consist of: (i) the flange plate, up to a maximum outstand width of $10 t_f \sqrt{\frac{355}{\sigma_{yf}}}$, where t_f is the thickness and σ_{yf} is the yield stress of the flange (this limit is meant to account for the torsional buckling of a wide flange outstand); (ii) an associated web plate depth of $12 t_w \sqrt{\frac{355}{\sigma_{yw}}}$ (this limit is meant to avoid using a substantial portion of the web to its full strength twice, i.e. yielding in diagonal tension and yielding due to formation of plastic hinge in the flange.

Furthermore, to avoid torsion on this idealised flange member, only a section symmetrical about the web midplane is taken as effective, and any portion of the flange outside this section of symmetry is ignored. When the two flanges are unequal, the value of m is conservatively taken as that of the smaller flange.

The ultimate shear capacity τ_u is not taken higher than τ_y, thus restricting the benefit of the frame mechanism of phase three within reasonable deformation limits for the whole girder. In the case of a beam with small flanges, the two plastic hinges in each flange coincide, i.e. in Fig. 5.22, $c = 0$; but some membrane action still occurs, hanging from the vertical stiffeners; it has been shown that putting $M_p = 0$ in equation (5.33) and maximising τ_u by differentiating with respect to θ_t leads to the true Basler solution described earlier. When the flanges are very substantial, $c = a$ and $\theta_t = 45°$; if the web is very thin as well, τ_{cr} or τ_l will be negligible, leading to the fully developed tension field suggested by Wagner for very thin webs. With wide spacing of vertical stiffeners, the benefit of the membrane action decreases as the angle θ_t decreases; thus the shear capacity falls, finally being no more than the elastic critical buckling value when the aspect ratio ϕ is infinity, i.e. the case of no intermediate vertical stiffeners.

5.4.6 Bending strength of thin-web girders

In a girder with a thin web subjected primarily to gradually increasing bending moment and restrained against lateral torsional buckling, at one stage local buckling of the compressive part of the web sets in. But this does not completely exhaust the bending capacity of the girder, i.e. the girder continues to resist further bending moment, or in other words it has post-buckling strength. Such a girder finally fails at an applied bending moment which is less than its plastic moment of resistance. At this stage, the compressive part of the web undergoes substantial buckling and consequently the compressive flange buckles into the web as shown in Fig. 5.23.

As a result of the buckling of the compressive part of the web, the distribution of bending stress changes from the ideal linear pattern, as shown in Fig. 5.23, and the web becomes less efficient. To quantify the reduction in the bending strength of the web, the following reduction factor was suggested by Cooper[9] for an I-beam of equal flanges and a web deeper than $5.7 t_w \sqrt{\dfrac{E}{\sigma_{yw}}}$:

$$1 - 0.0005 \frac{A_w}{A_f} \left(\frac{d}{t_w} - 5.7 \sqrt{\frac{E}{\sigma_{yw}}} \right) \tag{5.34}$$

where A_w and A_f are the area of the web and each flange, respectively, and d is the web depth. According to this approach there is no reduction in bending strength if $\dfrac{d}{t_w}$ is less than 137 and 165 for $\sigma_{yw} = 355$ N/mm² and 245 N/mm², respectively. Cooper's expression for reduction in bending strength can also be expressed as a reduced effective web thickness t_{we} as follows:

$$\frac{t_{we}}{t_w} = 1 - \left(\frac{d}{t_w} - 5.7 \sqrt{\frac{E}{\sigma_{yw}}} \right) \left(0.003 + 0.0005 \frac{A_w}{A_f} \right) \tag{5.35}$$

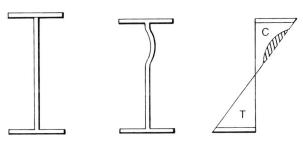

Fig. 5.23 Buckling of compressive part of web.

From the results of large-deflection elasto-plastic computer studies on the strength of plate panels subjected to in-plane bending and with different edge conditions, welding residual stresses and out-of-plane imperfections (see Section 5.4.4), the following expression for the effective width is proposed in BS 5400:Part 3[2]:

$$\frac{t_{we}}{t_w} = 1.425 - 0.00625 \frac{d_c}{t_w} \sqrt{\frac{\sigma_{yw}}{355}} \qquad (5.36)$$

where d_c is the depth of the compressive part of the web. This expression:

(1) ignores the effect of the different ratios of web to flange areas, as this effect, as predicted by equation (5.35), was in fact found to be quite small
(2) is valid for girders with equal or unequal flanges
(3) stipulates no reduction in the effectiveness of the web if the ratio of the depth of the compressive zone to thickness is less than 68 and 82 (or, in the case of equal flanges, the total web depth to its thickness ratio is less than 136 and 164) for steel yield stresses of 355 and 245 N/mm², respectively.

5.4.7 Combined bending moment and shear force on girders

Various researchers have extended their tension field models to apply to coexistent shear force and bending moment on the girder sections. For example, in the tension field model due to Rockey et al.[8], the elastic critical stress can be reduced to take into account interaction with bending moment by means of a formula like equation (5.25), and the plastic moment of resistance M_p of the flanges in equation (5.33) can be reduced by deducting the flange bending stress from its yield stress. In Basler's tension-field model, the shear is carried by the web only, without any contribution from the flange; but this shear capacity is only valid as long as the moment is less than the amount that can be resisted by the flanges only, i.e. less than $\sigma_{yf}A_f d$. Any larger moment must be resisted by bending stresses in the web which reduces its shear capacity, until the latter becomes very small, possibly zero, when the bending moment reaches the full plastic moment of the girder, or in the case of thin web girders its elastic moment of resistance. This interaction between moment and shear, in the absence of any lateral-torsional or web buckling, can be represented by Fig. 5.24(a), where V_u represents the full ultimate shear capacity of the girder and V_w represents the value ignoring any contribution from the flange.

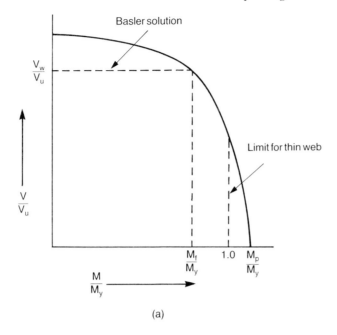

(a)

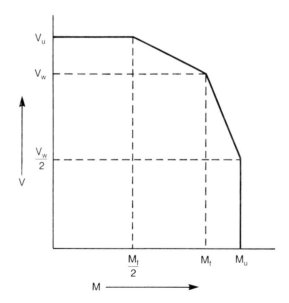

(b) As adopted in ref (2)

Fig. 5.24 Interaction between bending moment and shear.

In Reference 2 it has been assumed that, if the bending and shear capacities are calculated without any contribution from the web and the flange, respectively, then a beam can resist these values of bending moment M and shear force V even when coexistent; this has been verified from many test results. M_f is taken as $\sigma_b A_f d$ when σ_b is the limiting stress in the flange from equation (5.13), A_f is the flange area and d is the effective girder depth between its flanges; V_w is obtained by putting $M_p = 0$ in equation (5.32) or (5.33). It has been further assumed, supported by test results, that the bending capacity M_u of the girder as calculated for lateral-torsional buckling, elastic or plastic distribution of stresses or web buckling as appropriate, can be attained even when there is a coincident shear on the cross-section provided the latter is less than $\frac{1}{2}V_w$; similarly the full shear capacity V_u given by equation (5.33) is attained even in the presence of a coexistent bending moment provided the latter is less than $\frac{1}{2}M_f$. This interaction is shown in Fig. 5.24(b).

5.5   Design of stiffeners in plate girders

Stiffeners are often used to improve the buckling strength of plated structures. Deflection of the plate normal to its plane along the line of stiffening is resisted by the flexural rigidity of the stiffener. Buckling of stiffened plates occurs in two different modes, namely (i) the overall buckling in which the stiffeners buckle, and (ii) the local buckling in which the stiffeners remain stable forming nodal lines but the plate panels between the stiffeners buckle. As in the case of plate buckling, an ideally flat residual-stress-free stiffened panel initially remains flat till an elastic critical value of the loading is reached. Also, as in the case of plates, stiffened plates may often have substantial post-buckling strengths.

5.5.1   Elastic critical buckling of stiffened panels

Two basic methods are available for the analysis of elastic critical buckling of ideally flat residual-stress-free stiffened plates. In the classical method, either (i) a differential equation of equilibrium is solved in general terms by assuming a deflected shape and then the boundary conditions are used to obtain a characteristic equation for the elastic critical buckling load, or (ii) a Fourier-type series representation is set up for the possible deformation mode consistent with the boundary conditions and then an energy or work approach is applied. The second method, i.e. the numerical or computer-based method, can tackle the complex problems; in this method a solution is formed in terms of a discrete number of unknowns located at many points in the stiffened plate. Thus a large-order matrix equation is formed, whose coefficients are given by the geometry of the stiffened plate and its loading conditions; the elastic critical buckling load is the value of the load at which the determinant of the coefficients is zero. The critical buckling load of a stiffened panel is generally expressed as

$$\sigma_{cr} = k \frac{\pi^2 D}{b^2 t}$$

where k is a buckling coefficient that depends on the geometry of the stiffened plate, the loading pattern and the boundary conditions, plus three relative rigidities of the stiffener, given by:

- flexural: $\gamma = \dfrac{EI_s}{bd}$

- torsional: $\theta = \dfrac{GJ_s}{bd}$

- extensional: $\delta = \dfrac{A_s}{bt}$

and D is the flexural rigidity of the plate, equal to $\dfrac{Et^3}{12(1 - \mu^2)}$

 b is the spacing of the stiffeners
 t is the thickness of the plate
 I_s is the second moment of area of the stiffener cross-section, with the width of the plate acting with it, about its centroidal plane parallel to the plate
 J_s is the torsional constant of the stiffener cross-section
 A_s is the area of the cross-section.

Many solutions are available for a large range of stiffened panel geometries and loading types in References 10 and 11 for 'open'-type stiffener cross-sections (i.e. not 'closed'- or box-type cross-sections) with negligible torsional rigidity.

5.5.2   *The concept of optimum rigidity of stiffeners*

If the buckling stress σ_{cr} is plotted against the relative flexural rigidity γ for a stiffened plate with 'open'-type stiffeners, it can be found that initially σ_{cr} increases with γ; but after γ exceeds a certain value γ^*, there is no further increase in σ_{cr}. For $\gamma < \gamma^*$ the buckling of the stiffened plate involves bending of the stiffeners out of the plane of the plate as in mode (a) of Fig. 5.25, i.e. the overall buckling mode; but for $\gamma > \gamma^*$ the plate panels between the stiffeners and/or the boundaries buckle without any bending of the stiffeners, i.e. the stiffeners form the nodal lines of the buckling of the plate panels, as in mode (b) of Fig. 5.25, i.e. local buckling of plate panels. There is thus an optimum rigidity γ^* of the stiffeners for the maximum possible value of the buckling stress σ_{cr} of the whole stiffened panel (i.e. for overall buckling), which coincides with the elastic critical buckling stress of the individual plate panels of the stiffened plate (i.e. for local buckling). No further increase in the buckling load is possible by increasing γ beyond this optimum value.

The above concept of the optimum rigidity γ^* is thus based on the fundamental concept of the elastic critical buckling phenomenon of both the entire stiffened panel and the individual plate panels in it. As has been pointed out earlier, this concept is truly valid only for residual-stress-free perfectly flat plates with high yield stress; for a stiffened panel, in addition to this requirement the stiffeners must also be perfectly straight and residual-stress-free. In reality, just as plates have residual stresses and out-of-plane imperfections, so also the stiffeners have initial out-of-straightness and/or twist and residual stresses. As a result stiffeners tend to deflect even at low levels of applied loading. In the overall buckling mode, i.e. $\gamma < \gamma^*$, there is thus often no critical value of loading at which sudden buckling of the stiffeners occurs. Instead, as the applied load is gradually increased, the deflection of the stiffeners continues to increase at a gradually faster rate till no further increase in load can be resisted. This maximum load is usually less than the theoretical elastic critical value, but there are cases where, due to post-buckling reserve, the maximum load is higher than the latter.

In the local buckling mode, i.e. $\gamma > \gamma^*$, because of their initial crooked-ness, the stiffeners start deflecting even at low levels of loading and thus do not form non-deflecting nodal lines for the local buckling of the individual plate panels; as a consequence the elastic critical buckling load of the individual plate panels is often not reached even when they have very low residual stress and out-of-plane imperfections. The rigidity of the stiffeners has thus got to be n-times the theoretical optimum value γ^* in order to ensure buckling in mode (b), i.e. local buckling, of Fig. 5.25. The value of n depends upon the geometry of the stiffened panel and the type of loading and is found to vary from 2.5 to 5. Thus, according to the linear buckling theory of stiffened plates, either (i) stiffeners are provided with m-times the optimum rigidity γ^* that will ensure that the overall critical buckling stress is equal to the local critical buckling stress, or (ii) the value of the stiffener rigidity to be taken for calculating the overall critical

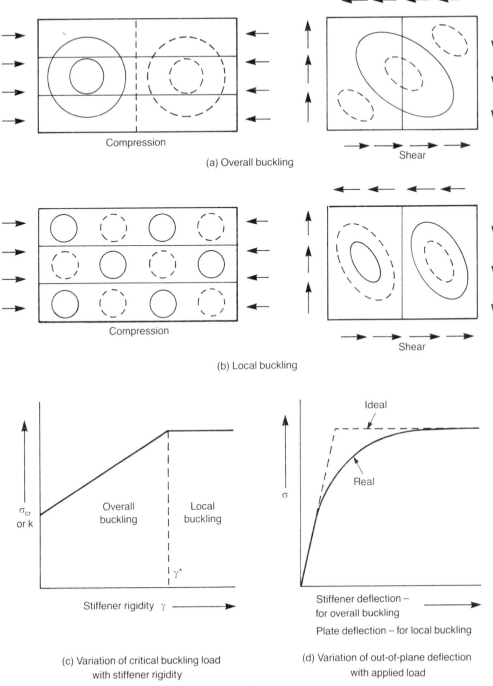

(a) Overall buckling

(b) Local buckling

(c) Variation of critical buckling load
with stiffener rigidity

(d) Variation of out-of-plane deflection
with applied load

Fig. 5.25   Buckling modes of stiffened panel.

buckling stress of the entire stiffened panel is obtained by dividing the actual rigidity value by *m*.

The above method for designing stiffeners has the following two serious shortfalls:

(1)   the exact value of the multiplying factor *m* is uncertain
(2)   it does not represent the behaviour of the web panels, and, more particularly, the compressive force in the vertical stiffeners, when a tension field is set up in the web.

5.5.3   Loading on a transverse web stiffener

The design of a transverse web stiffener should take account of the following effects if they are present:

(1)   the destabilising effect of the in-plane longitudinal and shear stresses in the web plate
(2)   axial force due to tension field in the web
(3)   axial force due to a locally applied concentrated load on the flange
(4)   axial force due to curvature or change of slope in the flange
(5)   axial force or bending moment transferred from a connected crossbeam or crossframe or deck.

In addition, a transverse stiffener at one end of a plate girder has also to resist the inward pull of the tension field in the plane of the web.

5.5.4   Destabilising effects of in-plane stresses in web

Stiffeners are provided to prevent the web plate from buckling due to the in-plane stresses in it. But when the loading is sufficiently increased the stiffeners themselves buckle. It may thus be assumed that the in-plane stresses in the web set up a bending tendency for the transverse stiffeners, which is resisted by their flexural stiffeners. This tendency can be visualised more clearly for a web subjected to longitudinal compressive stresses, as shown in Fig. 5.26. The elastic critical buckling stress of an orthogonally stiffened panel subjected to longitudinal compression is given by

$$\sigma_{cr} = \frac{\pi^2}{t_w a^2} \left[\frac{D_x}{\phi^2} + D_y \phi^2 + 2H \right]$$

where D_x and D_y are the flexural rigidities in the orthogonal directions, *H*

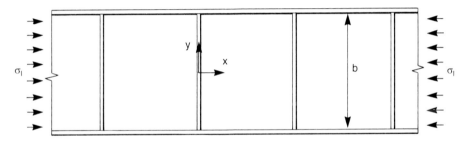

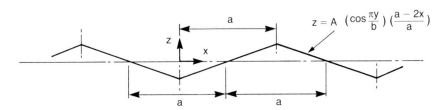

Fig. 5.26   Effect of longitudinal web stress on transverse stiffeners.

is the torsional rigidity of the panel, and ϕ is the aspect ratio $\dfrac{l}{b}$ of the buckled panel, l being the buckling half-wavelength in the x-direction.

In a web with only transverse stiffeners D_x and H are small compared with D_y, which is given by EI_s/a, i.e. the flexural stiffness of the transverse stiffeners smeared over their spacing a. Hence, for a minimum value of σ_{cr}, ϕ (and hence l) has to be a minimum. The smallest value of l that involves the buckling of the transverse stiffeners in any half-wave is obviously a, i.e. the spacing of these stiffeners. Since the rigidities D_x and H are very small, the flexure in the x-direction and torsional deformation in the x-y-plane may be ignored, and the buckled shape of the web may be assumed to be a saw-edge pattern in the x-direction with one transverse stiffener at each change of direction, as shown in Fig. 5.26. In the y-direction the shape may be assumed to be sinusoidal. σ_{cr} may be obtained by equating the strain energy of flexure of the transverse stiffener to the work done by stresses σ_1 due to shortening in the x-direction, or alternatively by solving the differential equation of flexure of the transverse stiffeners in terms of the stress σ_1. The strain energy of the stiffener can be shown to be

$$2 \int_o^{\frac{b}{2}} \frac{EI_s}{2} \left(\frac{d^2z}{dy^2}\right)^2 dy = \frac{\pi^4 \, EI_s \, A^2}{4b^3}$$

The average shortening of width a can be shown to be A^2/a, where A is the maximum out-of-plane deflection. Work done in shortening is then

$\sigma_1 t_w b A^2 / a$ and equating this with the strain energy, the critical value of σ_1 is obtained as

$$\sigma_{1cr} = \frac{\pi^4 \, EI_s \, a}{4 \, t_w b^4}$$

Alternatively, assume A_o and A_1 as the initial and the further amplitudes of the deflected shape of the stiffener. Then the load in the web causing the flexure of the stiffener is

$$\frac{4(A_o + A_1)}{a} \, \sigma_1 t_w \, \cos \frac{\pi y}{b}$$

which must be equal to

$$EI_s \frac{d^4 z}{dy^4} = EI_s \frac{\pi^4}{b^4} A_1 \cos \frac{\pi y}{b}$$

Equating,

$$A_1 = \frac{4 \, \sigma_1 t_w \, A_o}{\dfrac{\pi^4 \, EI_s \, a}{b^4} - 4 \, \sigma_1 t_w}$$

A_1 becomes infinity when

$$\sigma_1 = \frac{\pi^4 EI_s \, a}{4 \, t_w b^4}$$

which is thus the critical value σ_{1cr}.
   This can also be expressed as

$$\sigma_{1cr} = \frac{\pi^2 EI_s}{b^2} \frac{\pi^2 a}{4 \, t_w b^2}$$

The first term in the above expression is the Euler critical load of the stiffener as a strut. Hence the stiffener can be deemed to be a strut subjected to the destabilising effect of a load of

$$P_{eq} = \frac{4 \, \sigma_1 t_w b^2}{\pi^2 a} \qquad\qquad (5.37)$$

If the stiffener is subjected to a real axial load of P_a and the additional destabilising effect of P_{eq}, then the following modified Perry equation can be adopted for evaluating the limiting values of P_a and P_{eq} that will cause yielding in the extreme fibre:

$$P_a + (P_a + P_{eq}) \, \eta \, \frac{P_E}{P_E - (P_a + P_{eq})} = P_y \qquad\qquad (5.38)$$

where

$$P_E = \text{Euler buckling load } \frac{\pi^2 E I_s}{b^2}$$

$$P_y = \text{squash load} = \sigma_y A_s$$
$$A_s = \text{area of cross-section of stiffener}$$
$$\eta = \text{Perry imperfection parameter.}$$

If P_a is zero, then solving equation (5.38) for P_{eq} will give the limiting value of P acting alone on the stiffener.

Denoting this by P_{eql},

$$\frac{P_{eql} P_E \eta}{P_E - P_{eql}} = P_y$$

wherefrom

$$\frac{P_{eql}}{P_y} = \frac{1}{\eta + P_y/P_E} \tag{5.39}$$

The limiting value of P_a acting alone on the stiffener is the usual ultimate compressive load on a strut, say P_{al}.

The destabilising effect of the longitudinal stresses σ_l can thus be considered equivalent to an axial strut load of P_{ac} given by

$$P_{ac} = P_{eq} \frac{P_{al}}{P_{eql}}$$

$$= P_{eq} \frac{P_{al}}{P_y} \frac{P_y}{P_{eql}}$$

$$= \frac{4 \sigma_l t_w b^2}{\pi^2 a} \frac{P_{al}}{P_y} \left(\eta + \frac{P_y}{P_E} \right) \tag{5.40}$$

using equations (5.37) and (5.39).

From the above equation P_{ac} appears to be inversely proportional to the spacing a of the transverse stiffeners; this anomaly is due to the fact that the longitudinal flexural rigidity of the web plate has been ignored in the analysis. Hence, instead of the actual spacing a, the maximum spacing of the transverse stiffeners that will be permitted by the adopted web plate thickness may be used in equation (5.40).

A study of the elastic critical buckling solutions for many stiffened panel geometries in References 10 and 11 indicates that the magnitude of the critical shear stress of the panels is numerically very similar to the critical longitudinal compressive stress. Thus, σ_l above can be taken as the sum of the shear stress and the longitudinal stress in the web. When the distribution of longitudinal stress in the web is not purely uniform compression, but a combination of uniform compression σ_1 and pure in-plane longitudinal bending σ_b, such that $\sigma_1 \pm \sigma_b$ give the stresses at the edges of the web (see Fig. 5.10), then $(\sigma_1 + \frac{1}{6} \sigma_b)$ may be taken instead of σ_1 only in equation (5.40); this is based on the observation that the elastic critical buckling coefficient k is 4 for pure compression and 24 for pure bending (see Section 5.4.1).

5.5.5 *Axial compression due to tension field in web*

The tension field in the web constitutes diagonal tensile stresses in it, the vertical component of which has to be resisted by vertical web stiffeners. This is very like the force distribution in an N-type truss and is shown in Fig. 5.27.

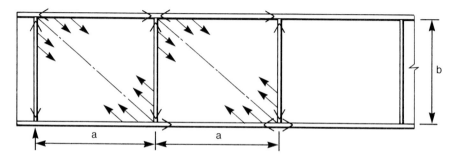

Fig. 5.27   Tension-field forces in a girder.

In Section 5.4.5 it has been postulated that when the applied shear stress exceeds the elastic critical value, the rest of the shear stress is resisted by the tension field mechanism. Under the combined action of shear and bending stresses in the web, elastic critical buckling occurs when the following condition is reached (see Section 5.4.1.4):

$$\frac{\sigma_1}{\sigma_{1cr}} + \left(\frac{\sigma_B}{\sigma_{Bcr}}\right)^2 + \left(\frac{\tau}{\tau_{cr}}\right)^2 = 1 \qquad (5.41)$$

where

σ_1 = uniform longitudinal compressive stress in web
σ_B = pure longitudinal bending stress in web
τ = shear stress in web

$$\left.\begin{array}{l}
\sigma_{1cr} = \dfrac{4\pi^2 E}{12(1 - \mu^2)} \left(\dfrac{t_w}{b}\right)^2 \\[12pt]
\sigma_{Bcr} = \dfrac{24\pi^2 E}{12(1 - \mu^2)} \left(\dfrac{t_w}{b}\right)^2 \\[12pt]
\tau_{cr} = \dfrac{k\pi^2 E}{12(1 - \mu^2)} \left(\dfrac{t_w}{b}\right)^2 \\[12pt]
k = 5.35 + 4 \left(\dfrac{b}{a}\right)^2
\end{array}\right\} \text{(see Section 5.4.1)}$$

Because of imperfections and residual stresses in the web, it may be conservatively assumed that tension field action starts when the applied stresses exceed 80% of the elastic critical buckling values, i.e. σ_{1cr}, σ_{Bcr} and τ_{cr} above are reduced by multiplying by 0.8.

A further simplifying and conservative step will be to assume the power of the second term in equation (5.41) to be 1, instead of 2. This will then lead to the limiting value of τ for the start of tension field action as

$$\tau_1 = 0.72Ek \left(\frac{t_w}{b}\right)^2 \sqrt{1 - \frac{\sigma_c}{2.9E} \left(\frac{b}{t_w}\right)^2} \tag{5.42}$$

where

$$\sigma_c = \sigma_1 + \tfrac{1}{6} \sigma_B$$

For the calculation of the forces on a vertical stiffener due to tension field in the web plate, it may be assumed that the membrane tensile stress is uniform over the whole web. The magnitude of this stress σ_t can be obtained by taking a vertical section through the web and equating the total vertical component of this stress to the shear force resisted by the tension field, as shown in Fig. 5.28.

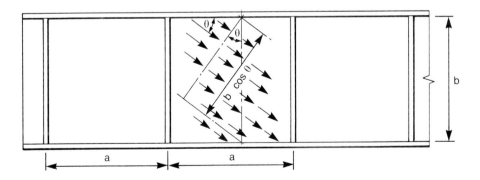

Fig. 5.28   Membrane forces due to tension field.

The force on the cut vertical section is given by

$$\sigma_t b \cos \theta_t t_w$$

and its vertical component is given by

$$(\sigma_t b \cos \theta_t t_w) \sin \theta_t$$

The shear force resisted by tension field action is

$$(\tau - \tau_l) \sigma t_w$$

Equating,

$$\sigma_t = \frac{(\tau - \tau_l)}{\sin \theta_t \cos \theta_t} \tag{5.43}$$

Due to σ_t, the pull on the flanges per unit length is

$$\sigma_t t_w \sin \theta_t$$

and its vertical component is

$$\sigma_t t_w \sin^2 \theta_t$$

The compressive force P_{tf} on a vertical stiffener is equal to the vertical component of the total pull over length a on the flanges. Thus,

$$P_{tf} = \sigma_t a t_w \sin^2 \theta_t$$

$$= \frac{\tau - \tau_l}{\sin \theta_t \cos \theta_t} a t_w \sin^2 \theta_t, \text{ using equation (5.43)}$$

$$= (\tau - \tau_l) a t_w \tan \theta_t \tag{5.44}$$

The inclination of the membrane forces θ_t for the maximum shear resistance due to tension-field action has been found from parametric studies never to exceed $\dfrac{\pi}{4}$, nor does it exceed the angle of the diagonal of the web panel with the horizontal for aspect ratio a/b of the panel up to 3. Hence P_{tf} can be taken as

$$P_{tf} = (\tau - \tau_l) a t_w \quad , \text{ or}$$

$$(\tau - \tau_l) b t_w$$

whichever is smaller.

Due to σ_t, the pull on an end vertical stiffener per unit height is

$$\sigma_t t_w \cos \theta_t$$

and its horizontal component is

$$\sigma_t t_w \cos^2 \theta_t$$

Assuming some end fixity, the bending moment on an end post is

$$\sigma_t t_w \cos^2 \theta_t \frac{b^2}{10}$$

$$= \frac{1}{10} (\tau - \tau_1) t_w b^2 \cot \theta_t$$

from equation (5.43).

From parametric studies for the maximum shear resistance due to tension-field action, an upper band for $\cot \theta_t$ was found to be $80/\theta_d$, where θ_d is the inclination of the diagonal of the web panel with the horizontal in degrees. The design bending moment M_y on an end stiffener can thus be expressed as

$$M_y = 8(\tau - \tau_1) t_w b^2 / \theta_d$$

5.6   Restraint at supports

The lateral buckling strength of beams has been derived in Section 5.3 on the assumption that its cross-section at the supports is restrained fully against any lateral deflection of its flanges. In reality the stiffeners at the support restraint to prevent twisting of the beam section is likely to be finite. Flint[12] has derived a reduction in the elastic critical bending strength σ_{cr} of a perfect simply supported beam as

$$\frac{\Delta \sigma_{cr}}{\sigma_{cr}}$$

$$= \frac{4}{3} \frac{GJ}{L_e S} \text{ for a central concentrated load}$$

$$= \frac{2GJ}{L_e S} \text{ for constant bending moment on the whole length of the beam.}$$

where GJ is the torsional rigidity of the beam
       L_e is the effective length of the beam
       S is the stiffness of the support against twisting of the beam section.

The reduction $\Delta \sigma_b$ in the limiting bending stress σ_b of the imperfect simply supported beam is, taking the worse of the above two cases,

$$\frac{\Delta \sigma_b}{\sigma_b} = \frac{\Delta \sigma_{cr}}{\sigma_{cr}} \frac{\Delta \sigma_b}{\Delta \sigma_{cr}} = \frac{2GJ}{L_e S} \frac{\Delta \sigma_b}{\Delta \sigma_{cr}}$$

If we decide to limit $\Delta\sigma_b$ to 1% of σ_b, then

$$\frac{2GJ}{L_eS}\frac{\Delta\sigma_b}{\Delta\sigma_{cr}} \not> 0.01$$

or

$$S \not< 200\frac{GJ}{L_e}\frac{\Delta\sigma_b}{\Delta\sigma_{cr}} \tag{5.45}$$

From equations (5.7) and (5.8) in Section 5.3, assuming k, v and the shape factor (i.e. the ratio of plastic to elastic modulus) to be approximately equal to unity, σ_{cr} can be expressed as

$$\frac{\sigma_{cr}}{\sigma_y} = \frac{\pi^2}{\left[\dfrac{L_e}{r_y}\sqrt{\dfrac{\sigma_y}{E}}\right]^2}$$

From equation (5.13), the limiting bending stress σ_b is given by

$$\frac{\sigma_b}{\sigma_y} = \frac{1}{2}\left\{1 + (1+\eta)\frac{\sigma_{cr}}{\sigma_y}\right\} - \frac{1}{2}\left[\left\{1 + (1+\eta)\frac{\sigma_{cr}}{\sigma_y}\right\}^2 - 4\frac{\sigma_{cr}}{\sigma_y}\right]^{\frac{1}{2}}$$

where, from (5.14),

$$\eta = 0, \text{ for } \frac{L_e}{r_y}\sqrt{\frac{\sigma_y}{E}} < 1.87$$

$$\eta = 0.12\left(\frac{L_e}{r_y}\sqrt{\frac{\sigma_y}{E}} - 1.87\right) \text{ for } \frac{L_e}{r_y}\sqrt{\frac{\sigma_y}{E}} > 1.87$$

r_y = radius of gyration about the vertical axis.

From the above two expressions of $\dfrac{\sigma_{cr}}{\sigma_y}$ and $\dfrac{\sigma_b}{\sigma_y}$, the derivative $\dfrac{\Delta\sigma_b}{\Delta\sigma_{cr}}$ has been found to be given approximately by

$$6 \times 10^{-3}\left(\frac{L_e}{r}\right) \quad \text{when } \frac{L_e}{r_y} > 100$$

$$6 \times 10^{-5}\left(\frac{L_e}{r_y}\right)^2 \quad \text{when } 50 < \frac{L_e}{r_y} < 100$$

$$1.2 \times 10^{-6}\left(\frac{L_e}{r_y}\right)^3 \quad \text{when } \frac{L_e}{r_y} < 50$$

Also, for I-section beams, the torsional stiffness GJ is approximately equal to

$$\frac{\pi^2}{2}\frac{EI_c D^2}{L^2}$$

where L is the simply supported span, D is the depth of the beam and I_c is the lateral moment of inertia of one flange.

I_c is approximately equal to 5 $r_y^3 T_c$, where T_c is the thickness of the compression flange. Using the above dimensional relationships in equation (5.44), the required stiffness of the support is given by

$$S \not< \alpha \ ET_c \ D^2$$

where

$$\alpha = 30 \left(\frac{r_y}{L_e}\right)^2 \text{ for } \frac{L_e}{r_y} > 100$$

$$= 0.3 \left(\frac{r_y}{L_e}\right) \text{ for } 50 < \frac{L_e}{r_y} < 100$$

$$= 0.006 \text{ for } \frac{L_e}{r_y} < 50$$

When the torsional restraint at the support section of a simply supported I-beam consists only of a load-bearing stiffener anchored to the pier or the abutment, the actual stiffness of the restraint is equal to

$$\frac{3EI_s}{D}$$

where I_s is the second moment of area of the cross-section of the load-bearing stiffener about a horizontal axis along the web of the beam.

5.7 In-plane restraint at flanges

In Section 5.4.4 it was stated that the behaviour and strength of an initially imperfect plate subject to in-plane loading depended upon the degree of in-plane restraint on the edges. In British Standard BS 5400: Part 3[2] separate graphs are given for the strength of:

(1) plates without any in-plane restraint on the longitudinal edges, i.e. free to pull in
(2) plates with longitudinal edges constrained to remain straight but allowed to move inwards in the plane.

Any in-plane restraint on the horizontal edges of an external web panel will apply a distributed load on the flange of a beam in the plane of the web; similarly, vertical edges at the ends of a beam will be subjected to in-plane pulling in forces for which a vertical stiffener over an external support needs to be designed. Along the internal edges of web plate panels the pulling-in forces in adjacent panels will tend to balance, and hence such edges may be deemed to remain virtually straight. For plates subjected to shear, its strength in the tension field mechanism depends upon the bending capacity of the flange (see Section 5.4.5). For plates subject to any pattern of in-plane loading, it is necessary to establish some criteria as to when the edges adjacent to a flange may be assumed to remain straight.

5.7.1   Plates under compression

For a plate panel under longitudinal compression σ_1, with initial imperfections in the critical buckling mode and with longitudinal edges held straight but free to pull in, the author has derived[14] the following expression for the transverse in-plane stresses at the longitudinal edges:

$$\sigma_2 = \frac{\pi^2}{8}(m^2 - 1)\frac{\delta^2 E}{b^2}\cos\frac{2\pi Nx}{L} \tag{5.46}$$

where L is the length of the plate in the longitudinal direction x

x is the longitudinal distance from the transverse centre line

N is the number of half-waves in direction x $\left(\approx \frac{L}{b}\right)$

b is the width of the plate

δ is the amplitude of the initial imperfection

m is the factor of magnification of initial imperfections under the loading σ_1

E is Young's modulus.

From the above expression, the maximum amplitude of σ_2 can be expressed as

$$\frac{\sigma_2}{\sigma_y} = \frac{\pi^3}{8}(m^2 - 1)\left(\frac{\delta_0}{S}\right)^2 \tag{5.47}$$

where δ_0 is a non-dimensional imperfection parameter δ/t

t is the plate thickness

S is a non-dimensional slenderness parameter $\dfrac{b}{t}\sqrt{\dfrac{\sigma_y}{E}}$

σ_y is the yield stress of the plate.

The author has also shown[14] that the value of m that is reached when the applied stress σ_1 reaches the ultimate strength of the panel is given by

$$0.78\delta_0^2 m^2 - \frac{1}{m} = 0.2766S^2 + 0.78\delta_0^2 - 1 \tag{5.48}$$

In Section 5.4.4 it was stated that, on the basis of the specified tolerances on workmanship, δ_0 can be assumed as 0.145 S. For any particular value of S, δ_0 can thus be quantified, and then from equation (5.48) the value of m may be obtained by successive approximation. The amplitude of the cosine distribution for σ_2/σ_y can then be obtained from equation (5.47). By this procedure a graph for σ_2/σ_y against slenderness parameter S was drawn. It was found that σ_2/σ_y could be taken as:

(1) zero, up to $S = 1.0$
(2) 0.5, for $S > 3.6$
(3) 0.192 $(S - 1)$, for $1.0 < S < 3.6$.

The cosine distribution of σ_2 on the edge, given by equation (5.46), causes a bending moment distribution on the girder flange on the edge, given by:

$$M = \frac{b^2}{4\pi^2} \sigma_2 t$$

The moment of resistance of the flange section M_R has to be at least a safety factor times the maximum value of the above bending moment. Taking account of the longitudinal stress already present in the flange due to the bending moment on the girder, the plastic moment of resistance M_R can be taken as

$$M_R = \frac{1}{4}\sigma_{yf} BT_f^2 \left[1 - \left(\frac{\sigma_f}{\sigma_{yf}}\right)^2 \right]$$

where B is the width of the flange
        T_f is the thickness of the flange
        σ_f is the longitudinal stress in the flange
        σ_{yf} is the yield stress of the flange.

Assuming a safety factor of 1.25,

$$M_R \geqslant 1.25M$$

or $$\frac{1}{4}\sigma_{yf} BT_f^2 \left[1 - \left(\frac{\sigma_f}{\sigma_{yf}}\right)^2 \right] \geqslant \frac{1.25b^2}{4\pi^2} \sigma_2 t$$

or $$\frac{\sigma_{yf} BT_f^2}{\sigma_y b^2 t} \geqslant \frac{1.25}{\pi^2} \frac{\sigma_2}{\sigma_y} \left(\frac{\sigma_{yf}^2}{\sigma_{yf}^2 - \sigma_f^2}\right)$$

For $S \leqslant 1.0$, σ_2 is zero and hence any web panel adjacent to a flange and with slenderness ratio $S \leqslant 1.0$ may be taken as restrained irrespective of the size of the girder flange. For web panels with $S \geqslant 3.6$, $\sigma_2 = 0.5\,\sigma_y$, and hence such a panel adjacent to a flange can be considered restrained only if the following condition is satisfied for the size of the flange:

$$\frac{\sigma_{yf} BT_f^2}{\sigma_y b^2 t} \geqslant 0.0625 \left(\frac{\sigma_{yf}^2}{\sigma_{yf}^2 - \sigma_f^2}\right) \qquad (5.49)$$

For a web panel adjacent to a flange and with slenderness ratio $1.0 < S < 3.6$, the flange size has to satisfy the following condition for the panel to be considered restrained:

$$\frac{\sigma_{yf}\, BT_f^2}{\sigma_y b^2 t} \geqslant \frac{1.25}{\pi^2}\, 0.192\, (S - 1) \left(\frac{\sigma_{yf}^2}{\sigma_{yf}^2 - \sigma_f^2}\right)$$

i.e.
$$\geqslant 0.024\, (S - 1) \left(\frac{\sigma_{yf}^2}{\sigma_{yf}^2 - \sigma_f^2}\right) \qquad (5.50)$$

5.7.2 Plates under shear

In Section 5.4.5 the tension-field strength of a plate subjected to shear was shown to depend upon a flange stiffness parameter m given by

$$m = \frac{M_p}{b^2 t\, \sigma_y}$$

This parameter is one-quarter of the ratio of the plastic moduli of the flange plate and the web plate about their respective horizontal centroidal axes, which has been used in the Section above for determining the flange stiffness required to achieve restrained plate capacity under compressive applied loading. The minimum values of m required to achieve in the tension field mechanism the levels of shear capacities represented by the restrained plate capacities in shear may be obtained from the expression for tension field strength for different aspect ratios of the plate. These minimum values should, however, be increased by a safety margin, since the tension field mechanism is attained with deformations in the plate, and in the flange member, of a much larger magnitude than is associated with the ultimate strength of orthogonally stiffened plates. By this method, the following stiffness requirement for the flange member has been derived as a condition for treating the web panels adjacent to the flange to be restrained to remain straight under in-plane shear loading:

$$\frac{\sigma_{yf}\, BT_f^2}{\sigma_y b^2 t} \geqslant 0.075\, (\phi)^{1.6} \left(\frac{\lambda - \bar{\lambda}}{200 - \bar{\lambda}}\right)^{0.3}, \text{ for } \lambda > \bar{\lambda} \qquad (5.51)$$

when $\lambda = \dfrac{b}{t} \sqrt{\dfrac{\sigma_y}{355}}$

$\bar{\lambda} = 66 + \dfrac{28}{\phi^2}$

$\phi = \text{aspect ratio } \dfrac{L}{b}$

(All other notation as in preceding Sections).

For $\lambda \leqslant \bar{\lambda}$ web panels may be deemed to be restrained, irrespective of the flexural stiffness of the flange.

5.8 Design example of a stiffened girder web

Web plate – 10 mm thick, grade 50 steel with $\sigma_y = 355$ N/mm². Applied longitudinal stresses due to factored dead and live loads are shown below. Applied shear stress = 60 N/mm². Further partial safety factors of $\gamma_{f3} = 1.1$ and $\gamma_m = 1.05$ are to be allowed for.

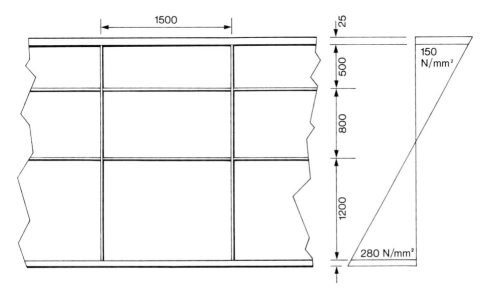

Fig. 5.29 Details of a design example of a girder web.

Yielding check – Hencky–Mises equivalent stress at top edge:

$$\sigma_e = \{150^2 + 3 \times 60^2\}^{0.5} = 182.5 \text{ N/mm}^2$$

Reference 2 allows for some plastic redistribution of the bending component of the longitudinal stress by taking only 0.77 times the bending stress. The top panel is subjected to 107 N/mm² direct compression and 43 N/mm² bending stress. Hence

$$\sigma_e = \{(107 + 0.77 \times 43)^2 + 3 \times 60^2\}^{0.5} = 176.4 \text{ N/mm}^2$$

Bottom edge:

$$\sigma_e = \{280^2 + 3 \times 60^2\}^{0.5} = 298.7 \text{ N/mm}^2$$

With plastic redistribution of bending stress in bottom panel,

$$\sigma_c = \{(176.8 + 0.77 \times 103.2)^2 + 3 \times 60^2\}^{0.5} = 276.5 \text{ N/mm}^2$$

Limiting value of

$$\sigma_c = 355/(1.1 \times 1.05) = 307.4 \text{ N/mm}^2$$

Hence the design is satisfactory for yielding.

Buckling check
(1) *Top panel*: to check if the panel can be deemed restrained for compressive and shear stresses. To take compression first,

$$\frac{BT_f^2}{b^2 t} = \frac{500 \times 25^2}{500^2 \times 10} = 0.125$$

which is greater than the minimum required as calculated below

$$\sigma_f = 152.15 \times 1.1 \times 1.05 = 175.7 \text{ N/mm}^2$$

$$S = \frac{b}{t} \sqrt{\frac{\sigma_y}{E}} = 50 \sqrt{\frac{355}{205\,000}} = 2.08$$

$$0.024 \, (S - 1) \, \{\sigma_{yf}^2/(\sigma_{yf}^2 - \sigma_f^2)\} = 0.024 \times 1.08 \times 1.325 = 0.0343$$

For shear, also, the slenderness ratio is such that the plate can be taken as restrained.

The panel is subjected to 107 N/mm² compressive, 43 N/mm² bending and 60 N/mm² shear stresses. Buckling stress coefficients are obtained from Reference 2 for

$$\frac{b}{t} \sqrt{\frac{\sigma_y}{355}} = 50 \text{ and } \phi = 3$$

as follows:

K_1 = coefficient for axial compression = 0.675
K_b = coefficient for pure bending   = 1.205
      (a value higher than 1.0 recognises plastic redistribution of bending
      stress at ultimate state)
K_q = coefficient for shear   = 0.966

For interaction of various stress components the following ratios are calculated:

$$m_c = \frac{\sigma_1 \, \gamma_m \, \gamma_{f3}}{\sigma_y \, K_1} = \frac{107 \times 1.05 \times 1.1}{355 \times 0.675} = 0.516$$

$$m_b = \left(\frac{\sigma_b \, \gamma_m \, \gamma_{f3}}{\sigma_y \, K_b}\right)^2 = \left(\frac{43 \times 1.05 \times 1.1}{355 \times 1.205}\right)^2 = 0.013$$

$$m_q = \left(\frac{\tau \, \gamma_m \, \gamma_{f3}}{\sigma_y \, K_q}\right)^2 = \left(\frac{60 \times 1.05 \times 1.1}{355 \times 0.966}\right)^2 = 0.041$$

$$m_c + m_b + 3m_q = 0.652 < 1.00$$

Hence the panel is safe for buckling.

(2) *Middle panel*: this panel is subjected to 4.8 N/mm^2 tensile, 68.8 N/mm^2 bending and 60 N/mm^2 shear stresses. Being an internal panel, it is deemed to be restrained.

Buckling stress coefficients are obtained from Reference 2 for

$$\frac{b}{t} \sqrt{\frac{\sigma_y}{355}} = 80, \text{ and } \phi = \frac{150}{80} = 1.875, \text{ as follows:}$$

$K_1 = 0.48$

$K_b = 1.15$, interpolating between $\phi = 1$ and 2

$K_q = 0.939$

$$m_c = \frac{4.8 \times 1.1 \times 1.05}{355 \times 0.48}$$

$$m_b = \left(\frac{68.8 \times 1.1 \times 1.05}{355 \times 1.15}\right)^2$$

$$m_q = \left(\frac{60 \times 1.1 \times 1.05}{355 \times 0.939}\right)^2$$

$$m_c + m_b + 3m_q \ll 1.00$$

Hence the panel is safe for buckling.

(3) *Bottom panel*: this panel is subjected to 176.8 N/mm^2 tensile, 103.2 N/mm^2 bending and 60 N/mm^2 shear stresses. First it has to be checked if the panel can be deemed to be restrained for shear and compression. Taking shear first,

$$\lambda = \frac{b}{t} \sqrt{\frac{\sigma_y}{355}} = 120; \quad \bar{\lambda} = 66 + \frac{28}{1.25^2} = 83.92$$

$$0.075(\phi)^{1.6} \left(\frac{\lambda - \bar{\lambda}}{200 - \bar{\lambda}}\right)^{0.3} = 0.075 \times 1.25^{1.6} \times \left(\frac{120 - 83.92}{200 - 83.92}\right)^{0.3}$$

$$= 0.0753$$

$$\frac{BT_f^2}{b^2 t} = \frac{500 \times 25^2}{1200^2 \times 10} = 0.0217, \text{ i.e. less than above.}$$

Hence the panel cannot be deemed restrained. Buckling coefficients for

$$\frac{b}{t} \sqrt{\frac{\sigma_y}{355}} = 120 \text{ and } \phi = 1.25$$

are obtained from Reference 2 as follows:

- Compression, $K_1 = 0.270$
- Bending, $K_b = 0.975$
- Shear, $K_q = 0.608$

$$m_c = \frac{-176.8 \times 1.1 \times 1.05}{0.27 \times 355} = -2.13$$

$$m_b = \left(\frac{103.2 \times 1.1 \times 1.05}{0.975 \times 355}\right)^2 = +0.119$$

$$m_q = \left(\frac{60 \times 1.1 \times 1.05}{0.608 \times 355}\right)^2 = +0.103$$

$$m_c + m_b + 3m_q \ll 1.00$$

Hence the panel is safe for buckling.

5.9 References

1. Standard Specification for Highway Bridges: 12th Edition: 1977: American Association of State Highway and Transportation Officials.
2. BS 5400: Part 3: 1982: Code of Practice for Design of Steel Bridges. British Standards Institution, London.
3. Kerensky O. A., Flint A. R. & Brown W. C. (1956) The basis for design of beams and plate girders in the revised British Standard 153. *Proceedings of the Institution of Civil Engineers* (August).
4. Johnston B. G. (Ed.) (1976) *Guide to Stability Design Criteria for Metal Structures*, Structural Stability Research Council. John Wiley & Sons.
5. Timoshenko S. P. & Gere J. M. (1961) *Theory of Elastic Stability*. McGraw-Hill Book Company.
6. Bulson P. S. (1970) *The Stability of Flat Plates*. Chatto and Windus.
7. Inquiry into the basis of design and method of erection of steel box girder bridges. In *Interim Design and Workmanship Rules*. HMSO, London (1973/4).
8. Porter D. M., Rockey K. C. and Evans H. R. (1975) The collapse behaviour of plate girders in shear. *The Structural Engineer* (August).
9. Cooper P. B. (1971) The ultimate bending moment for plate girders. In *IABSE Colloquium on Design of Plate and Box Girders for Ultimate Strength*, London.
10. Klöppel K. and Scheer J. *Beulwerte Ausgesteifter Rechteckplatten*. Vol. I. Wilhelm Ernst & Sohn, Berlin.
11. Klöppel K. and Möller K. H. *Beulwerte Ausgesteifter Rechteckplatten*. Vol II. Wilhelm Ernst & Sohn, Berlin.
12. Flint A. R. (1951) The influence of restraints on the stability of beams. *Structural Engineer* (September).
13. BS 5400: Part 4: 1979. Code of Practice for Design of Composite Bridges. British Standards Institution.
14. Chatterjee S. (1978) 'Ultimate load analysis and design of stiffened plates in compression.' PhD Thesis, London University.

Chapter 6
Stiffened Compression Flanges of Box and Plate Girders

6.1 General features

Stiffened compression flanges of box and plate girders consist of a flange plate stiffened longitudinally by either open (i.e. flat, bulb-flat, angle or tee) or closed (i.e. trough, vee) types of stiffeners spanning between transverse stiffeners which are supported by the girder webs or web stiffeners. Such compression flanges may be subjected to the following stresses:

(1) Longitudinal stresses due to the bending moment (and axial force) on the main girder; these stresses may vary across the width of the flange due to shear lag, and along the length due to the variation in the bending moment; additional longitudinal stresses may be caused by restrained warping of a box girder
(2) In-plane shear stresses in the flange plate due to any shear force on the girder and/or torsion in the case of a box girder
(3) Flexural stresses in the flange stiffeners due to any loading applied locally on the flange, e.g. wheel loading on a bridge deck
(4) In-plane transverse stresses in the flange plate due to bending of transverse flange stiffeners, and in the case of box girders due to distortion of the box cross-section and in the vicinity of internal diaphragms over box girder supports.

Typical stiffened flange details are shown in Fig. 6.1.

A stiffened compression flange comprises several parallel struts each continuous over and supported at many transverse stiffener locations; it can be idealised as a series of pin-ended struts supported at transverse stiffeners. Apart from the complex stress field mentioned above, the following geometrical complexities need investigation:

(1) Longitudinal continuity over transverse stiffeners
(2) Transverse continuity between parallel stiffeners

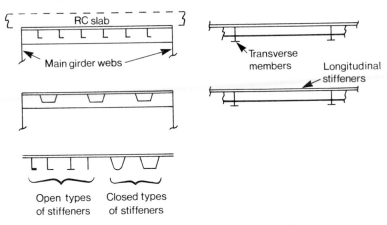

Fig. 6.1   Typical construction details of stiffened compression flange.

(3) Four separate buckling modes, namely local buckling of stiffener components, buckling of longitudinal stiffeners between transverse stiffeners, and overall buckling of the orthotropically stiffened panel between girder webs

(4) Geometric imperfections and residual stresses in the flange plate and the stiffeners.

   Any interaction between local buckling of a stiffener outstand and either local plate buckling or overall strut buckling would lead to sudden and substantial drop of load resistance. Hence design rules for stiffened plate structures specify such geometrical limitations on stiffener outstands that they are neither prone to premature buckling nor sensitive to any initial imperfections in them.

6.2   Buckling of flange plate

It will generally be uneconomical to limit the plate slenderness to such a value that it will not deform out-of-plane before the whole strut buckles. Therefore the flange plate may have a non-linear load-deformation response; and the flange plate being one component of the strut cross-section, this will affect the behaviour of the strut. The stress in the flange plate due to axial force P and bending moment M on the strut is given by

$$\sigma = k_s \left[\frac{P}{A_e} + \frac{My}{I_e} \right] \tag{6.1}$$

when k_s is the secant stiffness factor of the flange plate, appropriate to the value of stress σ in it, as shown in Fig. 6.2. A_e and I_e are the area and the second moment of the area of the strut cross-section in which k_s times the flange plate area is taken as effective, and y is the distance of the mid-plane of the flange plate from the centroidal axis of the effective strut section. k_s is given by $\dfrac{\sigma}{eE}$, where e is the longitudinal shortening per unit length of the flange plate corresponding to σ. For very stocky plates not liable to deform out-of-plane, eE equals σ and k_s equals unity. For slender plates, σ and k_s are interrelated and hence a method of successive approximation or trial-and-error is necessary for obtaining the stress σ in the flange plate due to any given applied loads and moments P and M on the strut. Obviously the stress σ must not exceed the ultimate strength σ_u of the plate shown in Fig. 6.2.

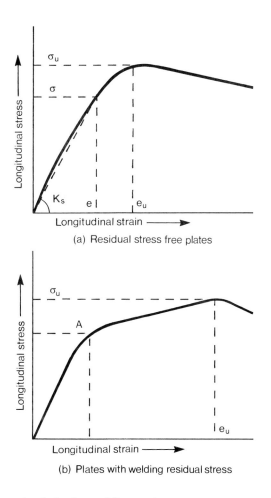

(a) Residual stress free plates

(b) Plates with welding residual stress

Fig. 6.2   Load-shortening behaviour of flange plates.

A rational design method for the flange stiffeners is thus:

(1)   To use a stiffness factor corresponding to σ_u, i.e. $k_s = \dfrac{\sigma_u}{e_u E}$, where e_u is the longitudinal strain at σ_u, for calculating the effective plate area, and

(2)   To limit the stress in the plate, calculated for the effective stiffener section by equation (6.1), to σ_u, i.e. σ_y times a strength factor $\dfrac{\sigma_u}{\sigma_y}$.

There are thus two necessary plate factors.

With a welding residual stress of 10% of the yield stress, and an initial imperfection amplitude of $\dfrac{1}{200}\sqrt{\dfrac{\sigma_y}{245}}$ times the smaller dimension of the plate (see Section 5.4.4), the above two stiffness factors corresponding to the initial linear limit A of Fig. 6.2(b) are shown in Fig. 6.3. However, it would be simpler to use only one plate effectiveness factor. For stocky welded plates, the ultimate strength actually reaches that of an unwelded plate, but at a higher longitudinal strain. Hence the strength effectiveness factor can be modified as shown in Fig. 6.3 and adopted as a single factor for both strength and stiffness. As a matter of fact, this modified factor became idential to the plate buckling factor in compression discussed in Section 5.4.4.

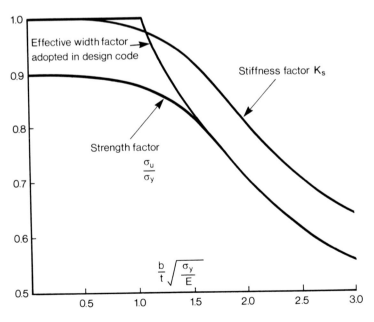

Fig. 6.3   Stiffness factors for welded plates.

6.3 Overall buckling of strut

The differential equation of equilibrium of an initially imperfect beam-column is

$$-EI_e \frac{d^2(y - y_o)}{dx^2} = P\delta$$

where δ is the eccentricity of the line of action of the axial load P with respect to the deflected centroid of the effective strut cross-section at any location in the strut length, y_o and y are the initial and the final positions of the centroid at that cross-section, and I_e is the second moment of area of the effective cross-section in which the secant stiffness factor times the nominal flange area is taken as effective.

The maximum bending moment M occurring at the mid-span of the length L of the strut can be shown to be very close to

$$M = P(e_1 + e_2) \frac{P_E}{P_E - P}$$

where P_E = Euler column buckling load $= \dfrac{\pi^2 EI_e}{L^2}$

     e_1 = end eccentricity of applied loading
     e_2 = maximum amplitude of a sinusoidal initial imperfection pattern
     L = length of the strut between its pinned ends.

The end eccentricity of applied loading on the flange stiffener arises from the non-uniformity of the stress distribution on the cross-section of a box or plate girder. It is theoretically reasonable to assume that the linear elastic theory of bending predicts satisfactorily the stress distribution on the girder cross-sections at the locations of stiff crossgirders or crossframes. Thus the magnitude and the pattern of the loading at the ends of the longitudinal flange stiffeners can be obtained by applying the simple beam theory to the girder cross-section. This loading pattern is shown in Fig. 6.4; the shaded block can be represented by an axial load $P = \sigma_a A_e$ acting at an eccentricity $e_1 = \dfrac{r^2}{h}$ with respect to the centroid of the stiffener cross-section, where r is the radius of gyration of the stiffener cross-section and h is the distance between the stiffener centroid and the neutral axis (i.e. level of zero stress) of the girder.

The well-known Perry strut equation can be used to obtain the limiting value of the longitudinal stress that can be applied on the effective strut:

$$\frac{\sigma_{su}}{\sigma_y'} = \frac{1}{2}\left[\left\{1 + (1 + \eta)\frac{\sigma_E}{\sigma_y'}\right\} - \sqrt{\left\{1 + (1 + \eta)\frac{\sigma_E}{\sigma_y'}\right\}^2 - \frac{4\,\sigma_E}{\sigma_y'}}\,\right] \qquad (6.2)$$

where σ_{su} is the limiting value of the applied longitudinal stress on the strut

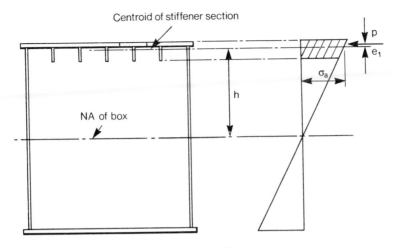

Fig. 6.4 Stress pattern on the ends of flange stiffeners.

σ_E is the Euler buckling stress of the strut

$$\eta = \frac{\Delta y}{r^2}$$

Δ is the maximum initial eccentricity and imperfection, i.e.
$(e_1 + e_2)$

y is the distance of the extreme compressive fibre from the centroid of the effective strut cross-section

r is the radius of gyration of the effective strut cross-section

σ_y' is the available yield stress at the extreme compressive fibre (see Section 6.4).

Because of the asymmetry of the cross-section about the horizontal centroidal axis, the above equation (6.2) must be applied to both the flange plate and the tip of the stiffening rib, with appropriate values for Δ, y and σ_y'.

6.4 Allowance for shear and transverse stress in flange plate

According to Hencky–Mises' criterion of yielding (see Chapter 2), the presence of shear stress in the flange plate reduces its effective yield stress to

$$\sigma_y' = \sqrt{\sigma_y^2 - 3\tau^2} \tag{6.3}$$

The flange shear stress τ is caused by (i) the vertical shear force on the cross-section of the main girder, and (ii) in the case of a box girder, by applied torsion on the girder. The stress due to (i) varies linearly from a maximum value over the main girder web to zero mid-way between a pair of such webs, and hence only half the maximum value needs to be taken along with the full value due to (ii) for τ in equation (6.3) for the flange plate initiated failure. For failure initiated by the tip of the stiffeners, the full value of the material yield stress of the tip is available in equation (6.2).

In addition to the above influence on yield stress, shear stress due to torsion on a box girder also causes a destabilising effect on the longitudinal flange stiffeners. An allowance for this may be made in the form of an additional notional axial load in the same way as derived for web stiffeners in Chapter 5.

Transverse stresses in the flange plate are caused by the flexure of the transverse flange stiffeners, and crossframes and diaphragms in box girders. As the centroid of the cross-section of a transverse stiffener is very near the flange plate, the magnitude of the transverse stress is small. When this transverse stress is compressive, it may in fact increase the effective yield stress in longitudinal compression of the longitudinal stiffeners, as per Hencky–Mises' yield criterion. Being localised, these transverse stresses are also not likely to cause any destabilising effects.

6.5   Orthotropic buckling of stiffened flange

Orthotropic buckling of the stiffened flange between the webs of the main girders provides a restraining effect on the buckling of the longitudinal flange stiffeners as individual parallel struts. This orthotropic behaviour produces two separate effects on the stress conditions of individual longitudinal stiffeners:

(1)   Under the same magnitude of the applied longitudinal compressive load across the flange width, the magnification of the initial deflection of the strut in an orthotropic panel is less than that of an isolated strut

(2)   As the whole flange width between girder webs buckles, the applied longitudinal compression varies across this width, with higher stresses along the edges (i.e. near the girder webs) and lower stresses along the central strips.

The word 'orthotropic' is an abbreviation of the feature 'orthogonally anisotropic'; in such a plate the lack of isotropy is due to different flexural rigidities in the orthogonal directions, even though the plate is of uniform thickness. The elastic critical buckling stress of an orthotropic plate under longitudinal compression, i.e. the value of the applied stress at which an ideally flat residual-stress-free orthotropic panel becomes unstable and suddenly deflects from its initially flat plane, is given by

$$\sigma_{cro} = \frac{\pi^2}{tb^2}\left[\frac{D_x}{\phi^2} + D_y\phi^2 + 2H\right]$$

(6.4)

where t is the thickness of the orthotropic plate

b is the width of the orthotropic plate

a is the length of the orthotropic plate

D_x, D_y are the flexural rigidities in the x and y directions, respectively

H is the torsional rigidity

ϕ is the aspect ratio of the buckled panel, i.e. $\dfrac{a}{mb}$

m is the number of half-waves in the longitudinal dimension a.

For a minimum value of σ_{cro}, $\dfrac{d\sigma_{cro}}{d\phi} = 0$, i.e. $\phi^4 = \dfrac{D_x}{D_y}$, leading to

$$\sigma_{cro} = \frac{2\pi^2}{tb^2}\left[\sqrt{D_xD_y} + H\right]$$

(6.5)

and the half-wavelength of buckling l in the longitudinal direction is given by

$$l = b\left[\frac{D_x}{D_y}\right]^{\frac{1}{4}}$$

This buckling mode of the whole stiffened flange will involve interactive buckling of both the longitudinal and the transverse stiffeners; this buckling mode will not only be sensitive to the initial imperfections of the stiffeners, but will be of a very 'brittle' nature, i.e. there will be a sudden and substantial shedding of the load at the onset of buckling. To avoid this catastrophic phenomenon, the transverse stiffeners are normally designed to be sufficiently stiff not to buckle when the longitudinal stiffeners do. Further reasons for making the transverse stiffeners sufficiently stiff are that on the top flange they have to support without large deflection any locally applied axle loadings of vehicles, and in box girders they form components of the internal cross-frames or diaphragms which are provided to prevent distortion of the box cross-section. With such rigid transverse stiffeners the buckling of the stiffened flange will have one half-wave in the longitudinal direction between adjacent transverse stiffeners. The rigidities D_x, D_y and H will thus not involve the geometric properties of the transverse stiffener and will be given by

$$\left.\begin{array}{l} D_x = \dfrac{EI_x}{b'} \qquad\qquad D_y = \dfrac{Et^3}{12(1 - v^2)} \\[3mm] H = \dfrac{Gt^3}{6} + \dfrac{1}{2}\,v_yD_x + \dfrac{1}{2}\,v_xD_y + \dfrac{GJ_x}{2b'} \end{array}\right\}$$

(6.6)

where I_x is the second moment of area of a longitudinal stiffener
b' is the spacing of longitudinal stiffeners
t is the flange plate thickness
v is Poisson's ratio of the flange plate
J_x is the torsional constant of the longitudinal stiffener

$$v_x = v \frac{b't}{b't + A_{sx}}$$

$$v_y = v$$

A_{sx} = cross-sectional area of one longitudinal stiffener.

For an ideal orthotropic plate, $v_y D_x = v_x D_y$. In the case of a compression flange discretely stiffened by longitudinal stiffeners between rigid transverse stiffeners, this equality is not satisfied and the contribution of D_x towards the torsional rigidity H is doubtful. Hence it is safe to take

$$H = \frac{Gt^3}{6} + v_x D_y + \frac{GJ_x}{2b'} \tag{6.7}$$

From equation (6.4), the total critical longitudinal compressive force in the whole orthotropic panel will be

$$\sigma_{cro} bt = \frac{\pi^2}{b} \left[\frac{D_x}{\phi^2} + D_y \phi^2 + 2H \right]$$

The real orthotropic flange has discrete flange stiffeners, each of cross-sectional area A_{sx}. If the width of the orthotropic flange panel between adjacent webs of main girders is denoted by B, instead of b, the average critical stress on the flange will be given by the above expression divided by $Bt + \Sigma A_{sx}$, i.e.

$$\sigma_{cro} = \frac{\pi^2}{B^2 \left(t + \frac{\Sigma A_{sx}}{B} \right)} \left[\frac{D_x}{\phi^2} + D_y \phi^2 + 2H \right]$$

The natural half-wavelength for the minimum value of σ_{cro} should be $B(D_x/D_y)^{\frac{1}{4}}$; but for the orthotropic panel buckling between adjacent transverse stiffeners, D_x and D_y are given in equations (6.6) and (D_x/D_y) will be very large. Hence the actual half-wavelengths will be limited to the spacing L between transverse stiffeners and ϕ should be taken as L/B. This leads to

$$\sigma_{cro} = \frac{\pi^2}{\left(t + \frac{\Sigma A_{sx}}{B} \right)} \left[\frac{D_x}{L^2} + \frac{D_y L^2}{B^4} + \frac{2H}{B^2} \right] \tag{6.8}$$

The minimum required stiffness I_y of transverse stiffeners, necessary to ensure that overall buckling involving the transverse stiffeners is not more critical than buckling between adjacent transverse stiffeners, may be obtained from equations (6.5) and (6.8). For this purpose we may note that in applying (6.5) we should take

$$D_y = \frac{EI_y}{L}, \qquad H = \frac{Gt^3}{6} + \frac{GJ_x}{2b'} + \frac{GJ_y}{2L}$$

and in applying (6.8) we should take

$$D_y = \frac{Et^3}{12(1 - v^2)}, \qquad H = \frac{Gt^3}{6} + v_x D_y + \frac{GJ_x}{2b'}$$

For torsionally weak longitudinal and transverse stiffeners, i.e. open-type stiffeners, H is negligible in applying (6.5), and both D_y and H are negligible in applying (6.8). The minimum value of I_y of transverse stiffeners for such stiffening, to ensure that overall buckling is less critical than buckling between transverse stiffeners, is given by

$$\frac{2}{B^2} \sqrt{D_x D_y} > \frac{D_x}{L^2}$$

leading to

$$I_y > \frac{B^4 I_x}{4b' L^3} \qquad (6.9)$$

This is only true for elastic critical buckling of an ideally flat stiffened panel. However, to prevent interactive buckling between local and overall modes, sudden drastic unloading and acute sensitivity to initial imperfections, I_y should be several times the above value.

For an isolated strut with a maximum initial out-of-straightness of e_1 in a sinusoidal mode, the magnification m of the out-of-straightness under an axial load P_a is given by

$$m = \frac{P_{cr}}{P_{cr} - P_a}$$

where P_{cr} is the Euler buckling load of the strut. The following cubic equation gives the magnification m of the initial imperfection/eccentricity Δ of an orthotropically stiffened panel of length L between adjacent rigid transverse stiffeners under the action of an applied compressive stress σ_a:

$$\sigma_a = \sigma_{cro} - \frac{\sigma_{cro}}{m} + \frac{E\Delta^2}{L^2} (m^2 - 1) \qquad (6.10)$$

where σ_{cro} is the elastic critical buckling compressive stress of the orthotropic panel and Δ is the sum of the maximum initial out-of-straightness in length L and any end eccentricity of the applied stress σ_a. The actual longitudinal stresses σ_c and σ_e along the longitudinal centre line and edges, respectively, of the orthotropic plate are given by:

$$\sigma_c = \sigma_a - \frac{2E\Delta^2}{L^2}(m^2 - 1) \left.\begin{array}{c}\\ \\ \\ \\ \\ \end{array}\right\}$$

$$\sigma_e = \sigma_a + \frac{2E\Delta^2}{L^2}(m^2 - 1)$$

(6.11)

A longitudinal stiffener along or near the centre line is thus subject to

(1)   an axial force $\sigma_c A_e$, where A_e is the effective stiffener cross-section
(2)   a maximum bending moment at its mid-span of

$$4\pi E I_e \Delta (m - 1)/L^2 \tag{6.12}$$

A stiffener at or near the longitudinal edge is subject to:

(1)   an axial force of $\sigma_e A_e$
(2)   a bending moment of $\sigma_e A_e \Delta$.

In an orthotropic stiffened panel, it is possible that all or most of the longitudinal stiffeners in the cross-section may have high initial imperfections e_1 of similar magnitude; the end eccentricity e_2 of the applied stress due to overall bending of the whole box and plate girder is also the same for all longitudinal stiffeners. The first Fourier series term for a constant value of $(e_1 + e_2)$ across the whole width of the cross-section is $\frac{4}{\pi}(e_1 + e_2)$, and equations (6.10) to (6.12) take account of this increase in the effective value of the imperfection.

The effective cross-section of a central stiffener and also an edge stiffener should be checked, with appropriate values and sign of Δ, so that the maximum stress due to the above longitudinal axial loads and bending moments does not exceed the effective yield stress of the flange plate given by equation (6.3) or the yield stress of the tip of the stiffener. The benefit of orthotropic action may be considerable in the cases of:

(1)   Narrow compression flanges between main girder webs, with only one or two longitudinal stiffeners
(2)   Shallow flange stiffeners, i.e. stiffeners with low Euler buckling stress
(3)   Closed type of flange stiffeners, e.g. troughs, that possess substantial torsional rigidity.

For stiffened flanges of the common types, i.e. with several open-type longitudinal stiffeners, with slenderness ratio l/r less than 60, the benefit from orthotropic behaviour is usually small. Another benefit of orthotropic behaviour in all stiffened flanges with reasonably stocky stiffener outstands is that, unlike isolated struts, the buckling behaviour at the ultimate load is of a stable nature, i.e. the load carried does not fall off sharply with increased longitudinal shortening.

6.6   Continuity of longitudinal stiffeners over transverse members

If the initial geometric imperfections of a continuous strut of several spans follow its critical buckling pattern, i.e. alternatively up and down in the successive spans, then the behaviour of the continuous strut under axial loading will be identical to that of a single-span strut. The end eccentricity of the applied longitudinal load is, however, in the same direction in all the spans and hence its effect on a continuous strut of several spans is different from that on a single-span strut; the bending moment caused by the end eccentricity will not only be reduced by the continuity, but will also vary from sagging to hogging in the various spans. Although the reduction in the magnitude of the moment is obviously helpful, the change of sign is important, owing to the asymmetry of the stiffener cross-section about the horizontal centroidal axis. The three-moment equation for axially loaded continuous beam columns of several spans, given in Reference 1, may be used to obtain the moments due to continuity which can then be combined with the moments from the simple span analysis. A coefficient F for effective eccentricity for any cross-section of a continuous beam column can be defined as:

$$F = \frac{\text{bending moment at a given section in a continuous beam column}}{\text{maximum bending moment in a simply supported beam column}}$$

For any particular section, F is not a constant but depends on the ratio of the applied axial load to the Euler critical buckling load. Values of F for mid-span and support sections for continuous beam columns of up to seven spans for axial load magnitude from zero up to 1.5 times the Euler buckling load varied from -0.50 to $+0.40$. The bending moments in the interior spans can be sagging or hogging — a feature which would not be detected by analysing a single-span pin-ended strut loaded eccentrically in a given direction. Hence the effective eccentricity of a continuous strut should be taken as $\pm$ half the actual eccentricity of the longitudinal loading.

6.7 Local transverse loading on stiffened compression flange

The three-moment theorem for continuous beam columns[1] is suitable for analysing a compression flange stiffener subjected to a local transverse loading as well as the axial compression due to the overall bending moment on the box or plate girder. In the case of an isolated simply supported column subjected to a transverse load P as well, the deflected form and hence the bending moments are found to be $P_E/(P_E - P)$ times those values had there been no axial loading, when P_E is the Euler critical load of an isolated column. In the analysis of a continuous beam column of several spans, the span moments are found to increase by the above factor, but the support moments are found to hardly increase. Provided the axial loading is less than half the Euler buckling loading of an isolated strut, the following method is appropriate for design:

(1) Calculate the bending moments due to local transverse loads on a continuous beam, ignoring the axial load
(2) To obtain the design span moments, multiply the values obtained in (1) by the factor $P_E/(P_E - P)$
(3) For design support moments, take the values obtained in (1).

Assume σ_a to be the axial stress on the effective strut section due to the applied axial load P, and σ_b to be the bending stress due to local transverse loading calculated without taking account of any axial loading. Then, with all the other terms as defined in Section 6.3:

(1) For the midspan region:

$$\sigma_a + \left[\frac{\sigma_a \Delta y}{r^2} + \sigma_b\right]\left[\frac{\sigma_E}{\sigma_E - \sigma_a}\right] \ngtr \sigma_y'$$

when σ_y' is the available yield stress at the extreme fibre (see Section 6.4).
    It can be shown that the above criterion can also be approximately expressed by the following simpler equation:

$$\frac{\sigma_a}{\sigma_{su}} + \frac{\sigma_b}{\sigma_y'} \ngtr 1 \qquad\qquad (6.13)$$

when σ_{su} is the strength of a strut without any local transverse loading, for which the effective material yield stress is σ_y' and as given by equation (6.2).
(2) For the support region:

$$\sigma_a + \left[\frac{\sigma_a \Delta y}{r^2}\right]\left[\frac{\sigma_E}{\sigma_E - \sigma_a}\right] + \sigma_b \ngtr \sigma_y'$$

or
$$\sigma_a + \left[\frac{\sigma_a \Delta y}{r^2}\right]\left[\frac{\sigma_E}{\sigma_E - \sigma_a}\right] \not> (\sigma_y' - \sigma_b)$$

or
$$\sigma_a \not> R\sigma_y'' \tag{6.14a}$$

where $R\sigma_y''$ is the strut strength of a strut made of material yield stress σ_y'':

$$\sigma_y'' = \sigma_y' - \sigma_b$$

and R is the ratio of the strut stength to material yield stress of a strut made of material yield stress σ_y''.

Alternatively,

$$\frac{\sigma_a}{R\sigma_y'} + \frac{\sigma_b}{\sigma_y'} \not> 1 \tag{6.14b}$$

Comparing equation (6.13) with this equation, it should be noted that σ_{su} is less than $R\sigma_y'$, as R is calculated for a strut of a lower effective material yield stress σ_y''; this reflects the benefit of the applied bending moment not being magnified by the axial load over the support section.

6.8 Effect of variation in the bending moment of a girder

The applied axial stress in a longitudinal stiffener in a stiffened flange varies according to the shape of the bending moment diagram on the box or plate girder. Such a stiffener may be checked as a uniformly compressed strut with an equivalent applied axial load equal to that occurring at a distance $0.4L$ from the heavily stressed end. However, for a relatively slender strut, i.e. one with high 'l/r' ratio, subjected to a substantial variation in the axial loading at its two ends, such an approximate method may be too conservative. A more accurate method, in which the magnitude and location of the maximum bending moment is calculated for such a strut with initial out-of-straightness imperfection, is given in Reference 2.

6.9 Transverse stiffeners in stiffened compression flanges

In the design of longitudinal stiffeners, it has been assumed that the transverse stiffeners shall provide adequate support to the longitudinal stiffeners, so that the effective buckling length of the latter does not exceed their span between the former, i.e. the transverse stiffeners form non-deflecting nodal lines in the buckling mode of the whole stiffened flange. Comparing the elastic critical buckling stresses for the buckling mode between adjacent transverse stiffeners and that involving their deflection, the maximum flexural rigidity of the transverse stiffener has been derived in equation (6.9) for a stiffened panel of 'open' type, i.e. torsionally weak stiffeners. This approach may be: (i) too conservative in some cases, as the flexural stiffness of the longitudinal stiffeners may be unnecessarily larger than that required for the longitudinal loading they carry; and (ii) too optimistic in other cases, as interactive buckling between the local mode (i.e. between transverse stiffeners) and the overall mode (i.e. involving deflection of transverse stiffeners) may produce a brittle type of failure with sudden and substantial fall-off in the applied loading. A more rational approach is to provide sufficient flexural rigidity of the transverse stiffeners so that there is a substantial safety factor over the magnitude of the applied loading against overall elastic critical buckling involving the deflection of transverse stiffeners.

The elastic critical buckling stress given by equation (6.4) has been derived for one isolated orthotropically stiffened panel, with the longitudinal edges simply supported. The overall elastic critical buckling stress of the entire width of a stiffened compression flange supported by several main girder webs and thus comprising several internal panels and one or two cantilever panels, is complex. There will be some interaction between adjacent panels if they are of different widths, through some rotational restraint at the common edge, but the magnitude of such interaction at the ultimate limit state is uncertain. For those reasons it is advisable to divide the whole width of the stiffened flange into a number of independent longitudinal panels and to calculate the critical buckling stress separately for each panel, i.e. assuming no interaction between adjacent panels. A cantilever panel on its own, i.e. without any rotational restraint at the supported longitudinal edge, tends to buckle in a mode which remains virtually straight in the transverse direction at all transverse sections and with very long half-wavelengths in the longitudinal direction; as a consequence the elastic critical buckling load is found to be almost entirely contributed by the torsional rigidity of the longitudinal stiffeners. For a stiffened flange with torsionally weak, i.e. 'open', type longitudinal stiffeners, the cantilever panels are virtually unstable unless they are combined with the adjacent panel or panels with both longitudinal edges simply supported. There are thus three types of deck panels to be considered, namely a deck panel without any cantilevers, a deck panel with one cantilever, and a deck panel with two cantilevers. For a stiffened flange with torsionally weak longitudinal stiffeners, the elastic critical buckling load per unit width of the stiffened flange is given by

$$p_{cr} = \frac{4}{B^2} \left[\frac{E_c I_c E_f I_f Y}{L} \right]^{\frac{1}{2}} \tag{6.15}$$

where $E_c I_c$ is the flexural rigidity of the transverse stiffener between the main girder webs, $E_f I_f$ is the flexural rigidity of the stiffened flange per unit width between the main girder webs, B is the span of the transverse stiffener between the main girder webs, L is the spacing of the transverse stiffener, and Y is a buckling coefficient depending on the various dimensions of the stiffened panel between the main girder webs and the cantilever portion.

Y is 24 for a stiffened panel between main girder webs without any cantilever overhangs. With a cantilever overhang on one or both sides the value of Y depends on the following two ratios:

(1) $\dfrac{I_{cc}}{I_c}$, where I_{cc} is the second moment of area of the cantilever portion

of the transverse stiffener and I_c that of the portion between the main girder webs

(2) $\dfrac{B_c}{B}$, where B_c is the width of the cantilever and B is the

width between the main girder webs.

Values of Y for different values of the above ratios are given in Table 6.1; the first set of figures applies to the case of the cantilever on one side only, and the figures in brackets apply to the case of the cantilever on both sides of an intermediate panel (see Fig. 6.5).

A continuous edge stiffening member may increase the critical buckling load of the stiffened panel because of the former's flexural stiffness, but the compressive force in it acting furthest from the support line of the panel also enhances the latter's buckling tendency. The transition takes

Table 6.1 Coefficients for overall buckling of stiffened flange with cantilever(s).

$\dfrac{B_c}{B}$	$\dfrac{I_{cc}}{I_c}$			
	0.2	0.4	0.6	0.8
0.2	22.9	22.9	22.9	22.9
	(21.9)	(21.9)	(21.9)	(21.9)
0.4	10.8	13.3	14.3	14.8
	(8.6)	(10.5)	(11.2)	(11.5)
0.6	2.7	4.2	5.1	5.6
	(2.3)	(3.3)	(3.9)	(4.2)
0.8	0.9	1.5	1.9	2.2
	(0.8)	(1.2)	(1.5)	(1.7)

Note: First figures apply to stiffened panels with cantilever on one side and figures in brackets apply to stiffened panels with cantilevers on both sides.

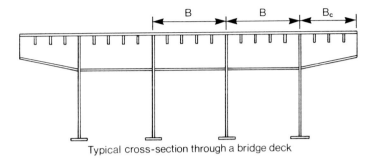

Typical cross-section through a bridge deck

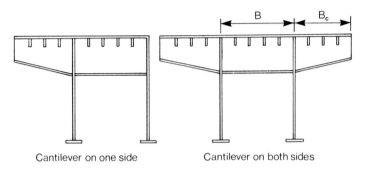

Cantilever on one side   Cantilever on both sides

Fig. 6.5   Geometry of stiffened compression flange.

place if the radius of gyration of the edge member is about 1.65 times that of the stiffened flange.

A more comprehensive expression for the critical buckling load of stiffened flanges outside the limitations stated above is given in Reference 3.

If σ_a is the applied longitudinal stress on the stiffened flange and A_f is its area per unit width, then a safety factor of 3 against elastic critical buckling involving deflection of transverse members will require the following minimum value for the transverse members:

$$\text{Minimum } I_c = \frac{9}{16Y}\frac{\sigma_a^2 A_f^2 L B^4}{E_c E_f I_f}$$

6.10   Stiffened compression flange without transverse stiffeners

Construction of a narrow box girder of moderate size may be considerably simplified if only one or two longitudinal stiffeners are provided on the compression flange, without any other form of stiffening. In such a design the buckling of the compression flange will be completely governed by the orthotropic behaviour of the stiffened flange panel. From Section 6.5 the critical buckling stress is given by

$$\sigma_{cro} = \frac{\pi^2}{B^2 \left(t + \frac{\Sigma A_{sx}}{B} \right)} \left[\frac{D_x}{\phi^2} + D_y \phi^2 + 2H \right]$$

where all the notation is as given in that Section. If the contribution of H is ignored, the expression for σ_{cro} can be reduced to

$$\sigma_{cro} = \frac{2\pi^2}{B^2 \left(t + \frac{\Sigma A_{sx}}{B} \right)} \frac{D_x}{\phi^2} = \frac{2\pi^2 E I_x}{l^2 b' t_e} \tag{6.16}$$

where $\qquad l$ = buckling half-wavelength = $B(D_x/D_y)^{0.25}$

$$\phi = \frac{l}{B}$$

t_e = effective thickness = $t + \frac{\Sigma A_{sx}}{B}$

B = width of flange between webs

I_x = moment of inertia of one flange stiffener

b' = spacing between stiffeners.

The above expression shows that an individual flange stiffener can be deemed to be an isolated Euler strut with an effective length L_e given by

$$L_e = \frac{l}{\sqrt{2}}$$

$$= \frac{B}{\sqrt{2}} \left(\frac{D_x}{D_y} \right)^{0.25}$$

We know that

$$D_x = \frac{E I_x}{b'} = \frac{E I_x (N + 1)}{B}$$

and

$$D_y = \frac{E t^3}{12(1 - v^2)}$$

where $\qquad N$ = number of flange stiffeners in flange width B

v = Poisson's ratio = 0.3.

Hence

$$L_e = \frac{B}{\sqrt{2}} \left[\frac{12 I_x (N + 1)(1 - v^2)}{B t^3} \right]^{0.25}$$

$$= 1.285 \left(\frac{B}{t} \right)^{0.75} \left[I_x (N + 1) \right]^{0.25}$$

To be on the conservative side, it is advisable to take

$$L = 1.5\left(\frac{B}{t}\right)^{0.25}\left[I_x(N+1)\right]^{0.25} \qquad (6.17)$$

6.11   A design example of stiffened compression flange

Applied longitudinal stresses due to factored loads are shown below; further partial safety factors of $\gamma_{f3} = 1.1$ and $\gamma_m = 1.20$ are to be allowed for.

A flange stiffener can be taken as a strut composed of a flange plate 375×25 and an angle $200 \times 100 \times 12$ (Fig. 6.6).

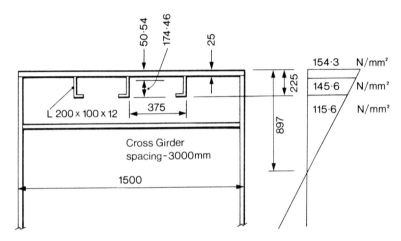

Fig. 6.6   Details of a design example of a stiffened compression flange.

The geometrical properties of this strut are:

Centroidal moment of inertia, $I_x = 64\,840\,298$ mm^4
area of cross-section = $12\,831$ mm^2
radius of gyration, $r =$ 71.09 mm
Maximum fibre distance to the top = 50.54 mm
Maximum fibre distance to the bottom = 174.46 mm

$e_1 =$ effective end eccentricity of applied loading $= \dfrac{1}{2}\dfrac{(71.09)^2}{897} = 2.82$ mm

$e_2 =$ amplitude of initial imperfection $= \dfrac{3000}{625} = 4.80$ mm

$$\Delta_1 = e_1 + e_2 = 7.62 \text{ mm}$$

$$\eta = \frac{7.62 \times 174.46}{(71.09)^2} = 0.263$$

Applied stress at the centroid $= 145.6$ N/mm^2

Euler buckling stress $\sigma_E = \pi^2 \times 20\,500 \times (71.09)^2/3000^2 = 1136$ N/mm^2

The ultimate stress σ_{su} is given by

$$\frac{\sigma_{su}}{355} = \frac{1}{2}\left[\left\{1 + 1.263\,\frac{1136}{355}\right\} - \sqrt{\left\{1 + 1.263\,\frac{1136}{355}\right\}^2 - \frac{4 \times 1136}{355}}\,\right]$$

$$= \tfrac{1}{2}[5.0416 - 3.552]$$

$$= 0.745$$

$$\sigma_{su} = 264.5 \text{ N/mm}^2$$

$$\frac{\sigma_{su}}{\gamma_m \gamma_{f3}} = \frac{264.5}{1.20 \times 1.1} = 200.4 \text{ N/mm}^2 > 145.6 \text{ N/mm}^2$$

Hence the flange stiffener is satisfactory.

If orthotropic behaviour of the whole flange is taken into account:

$$D_x = \frac{E \times 64\,840\,298}{375} = 172\,907E$$

$$D_y = \frac{E \times 25^3}{12 \times 0.91} = 1431E$$

$$H = 0.4E\,\frac{25^3}{6} + 0.3\,\frac{375 \times 25}{375 \times 25 + 3456} \times 1431E$$

$$= (1042 + 313)E = 1355E$$

$$\sigma_{cro} = \frac{\pi^2 E}{25 + \dfrac{3456 \times 3}{1500}}\left[\frac{172\,907}{3000^2} + \frac{1431 \times 3000^2}{1500^4} + \frac{2 \times 1355}{1500^2}\right]$$

$$= \frac{\pi^2 \times 205\,000}{25 + 6.912}\left[0.0192 + 0.0025 + 0.0012\right]$$

$$= 1452 \text{ N/mm}^2$$

$$\Delta = 7.62 \text{ mm}$$

$$\sigma_a = 145.6 \text{ N/mm}^2$$

The magnification of initial imperfection Δ under the applied stress σ_a is given by the relationship

$$\sigma_a = \sigma_{cro} - \frac{\sigma_{cro}}{m} + \frac{E\Delta^2}{L^2}(m^2 - 1)$$

Since this is a cubic equation, a method of successive approximation could be used. The first trial value for m is

$$m_1 = \frac{\sigma_{cro}}{\sigma_{cro} - \sigma_a} = \frac{1452}{1452 - 145.6} = 1.111$$

The corresponding applied stress σ_{a1} is calculated from the above relationship as 145.4 N/mm^2:

Derivative

$$\frac{d\sigma_a}{dm} = \frac{\sigma_{cro}}{m_1^2} + \frac{2E\Delta^2 m_1}{L^2} = 1179.3$$

The next approximation for m is

$$m_2 = m_1 - \frac{\sigma_{a1} - \sigma_a}{\left(\dfrac{d\sigma_o}{dm}\right)} = 1.1111$$

The longitudinal stress along a central stiffener is

$$\sigma_c = \sigma_a - \frac{2E\Delta^2}{L^2}(m^2 - 1) = 145.0 \text{ N/mm}^2$$

The longitudinal stress along a stiffener near the box edge is

$$\sigma_e = \sigma_a + \frac{2E\Delta^2}{L^2}(m^2 - 1) = 146.2 \text{ N/mm}^2$$

Central stiffener:

Bending moment $= 4\pi E I_e \Delta (m - 1)/L^2$

$$= \frac{4\pi \times 205\,000 \times I_e \times 7.621 \times 0.1111}{3000^2}$$

$$= 0.242 I_e$$

Max stress $= 145.0 + \dfrac{0.242 I_e \times 50.54}{I_e}$

$$= 145.0 + 12.2 = 157.2 \text{ N/mm}^2$$

Outer stiffener:

Bending moment $= 146.2 \times 12\,831 \times 7.62$ N.mm

Max stress $= 146.2 + \dfrac{146.2 \times 12\,831 \times 7.62}{64\,840\,298} \times 50.54$

$$= 146.2 + 11.1 = 157.3 \text{ N/mm}^2$$

These magnitudes are less than

$$\frac{\sigma_y}{\gamma_m \gamma_{f3}} = \frac{355}{1.20 \times 1.1} = 268.9 \text{ N/mm}^2$$

Hence the compression flange design is satisfactory.

6.12   References

1. Timoshenko S. P. and Gere J. M. (1961) *Theory of Elastic Stability*. McGraw-Hill Book Company.
2. Chatterjee S. (1978) *Ultimate load analysis and design of stiffened plates in compression*. Ph.D thesis, London University.
3. BS 5400: Part 3: 1982 Code of Practice for Design of Steel Bridges. British Standards Institution, London.

Index